Flexible Adaptation in Cognitive Radios

ANALOG CIRCUITS AND SIGNAL PROCESSING

Series Editors:
Mohammed Ismail. The Ohio State University
Mohamad Sawan. École Polytechnique de Montréal

For further volumes:
http://www.springer.com/series/7381

Shujun Li • Mieczyslaw M. Kokar

Flexible Adaptation in Cognitive Radios

 Springer

Shujun Li
Northeastern University
Boston
USA

Mieczyslaw M. Kokar
Northeastern University
Boston
USA

ISBN 978-1-4899-8866-9   ISBN 978-1-4614-0968-7 (eBook)
DOI 10.1007/978-1-4614-0968-7
Springer New York Heidelberg Dordrecht London

Printed on acid-free paper

Springer is part of Springer Science+Business Media (www.springer.com)

To Mom, Dad, Xu, and Winston

Preface

The concept of *cognitive radio*, first introduced by Mitola in 2000, envisioned it as a cognitive agent capable of autonomous operation in pursuing its own goals, communication with other agents, flexible adaptation, reasoning and learning. However, nowadays the term "cognitive radio" is typically interpreted as a radio that is able to monitor the unused spectrum in its environment and selecting spectrum for own transmissions without interfering with the licensed users of the given spectrum band.

This book takes up on the challenges of realizing the original goals of cognitive radio. Although it focuses on the topic of flexible link adaptation, the various technical solutions presented in this book will be useful for realizing the rest of the goals of the original vision for cognitive radio.

While adaptation is incorporated in essentially all communications protocols, such adaptations are limited to the situations envisioned by the protocol and radio designers. Protocol-based solutions cannot deal with the unknown. In other words, such adaptations lack the flexibility in the types of feedback they can process, the types of responses to the feedback they can invoke and most importantly they cannot negotiate better solutions with other radios or network nodes.

A greater flexibility of adaptation could be achieved via two mechanisms—collaborative negotiation of solutions and learning. In this book we are focusing on the first of the approaches. In communications, like in dancing, it takes two to the tango. One radio cannot modify the communications protocol without agreeing on such a change with its peers. Neither can it dictate what the peer needs to do without considering the capabilities of the other radio. Radios need to negotiate changes in the protocol. The question is—what language can they use for such negotiations. This book describes such a language and mechanisms to achieve collaboration. The potential usability and power of such a language should not be underestimated.

Boston, MA   Shujun Li and Mieczyslaw M. Kokar

Acknowledgement

Five years ago, back in September 2007, I started my Ph.D program with little clue of which research direction I am heading for and what my life will be in the following few years. One day, a friend in our department told me that Prof. Kokar is looking for a Research Assistant. After browsing his personal website, I decided to give it a try. Everything went very smoothly and two weeks later I became Prof. Kokar's student. I didn't realize that how lucky I am to make such a decision, a decision that will pave the way for a 4-year adventure full of joyful challenge and fruitful harvest.

This book is a memento of the great days in these 4 years. I would like to express my deepest gratitude to Prof. Mieczyslaw M. Kokar, my Ph.D advisor. His passion, perseverance and patience always inspired me with new ideas and encouraged me not to give up in the difficult times. His humor always spiced up the lengthy discussions we had in his office and also made him a good company during our trips to numerous academic conferences throughout the years. Without his constant guidance, advice and support, this book would not exist. I am also enormously grateful to my colleague Dr. Jakub Moskal for being such a warm and supportive friend and for his help on the implementation of the use case. I owe big thanks to those in the Wireless Innovation Forum and IEEE P1900.5 Working Group who inspired me with new ideas and gave detailed comments on my work through the face-to-face meetings, teleconferences and email discussions.

I would like to thank my editor Dr. Mohammed Ismail. He attended my presentation at the SDR'10 conference and encouraged me to publish my thesis into this book. I also want to thank my editors Charles Glaser and Augustine Leo for their patience and helpfulness through the book production process.

I give my special thanks to my parents, Yujuan Xuan and Jianping Li, and my uncle Wenzhan Xuan, for their unconditional love and support in the past thirty years.

Today marks the 11th anniversary of my husband Xu Ning and I being together. No word can express how grateful I am to have him in my life. Throughout the years, his love, tolerance, passion for technology and curiosity about the world has always fueled me to achieve my fullest potential and pursue those unimaginable dreams. Finally, I want to dedicate this book as the first gift to our baby Winston, who is expected to finally meet us in a few weeks.

Contents

List of Tables

List of Figures

Chapter 1
Introduction

1.1 Cognitive Radio and Dynamic Spectrum Access

In the wireless communication community the term "cognitive radio" (CR) is usually associated with the capability of *dynamic spectrum access* (DSA). This means the capability of a radio to utilize some areas of the electromagnetic spectrum that currently are not occupied by other radios, without causing interference problems to other radios. The increased interest in the development of new radio technologies to address the emergence of the shortage of spectrum is well justified economically and is supported by the various studies of spectrum utilization.

However, at the same time, the precious spectrum resource is highly underutilized. For example, it has been shown that on average, less than 5 %, and possibly as little as 1 %, of the spectrum below 3 GHz, as measured in frequency–space–time, is used (Kolodzy 2009). Also, there is spectrum that is never accessed or accessed only for a fraction of time.

Since radio spectrum is such a precious and expensive resource that has value not only to particular radio users, but also has a strategic value to the society as a whole, it is subject to both aggressive business competition and government regulation. In particular, in the United States, licenses for the use of the particular bands of the electromagnetic spectrum are auctioned by the Federal Communications Commission (FCC). The auctions are a source of revenue for the government. It is estimated that those auctions brought the revenue of over $60 billion to the US Government. One of the events that spurred an increased interest in the use of spectrum was the transition from analog to digital broadcast of TV signals in June of 2009. The 700 MHz spectrum became available (known as the "white spaces") and was partially auctioned. It has been reported that the price for this spectrum was $200 M/MHz (Fette 2009).

The cognitive radio technology has thus been recognized as one of the technologies that would solve the spectrum shortage problem. One of the main features of cognitive radio technology is the capability of sensing the environment. In this approach, a cognitive radio enables opportunistic spectrum access by sensing the

S. Li and M.M. Kokar, *Flexible Adaptation in Cognitive Radios*,
Analog Circuits and Signal Processing, DOI 10.1007/978-1-4614-0968-7_1,
© Springer Science+Business Media, LLC 2013

environment, detecting the underutilized spectrum at a specific time and location, and then adjusting the radio's transmission parameters to conform to the opportunity without harmful degradation to the primary user (Mitola and G. Q. Maguire 1999; Cognitive Radio Working Group 2008; Tsui 2004; Giacomoni and Sicker 2005; Radiocommunication Study Groups 2011; Haykin 2005).

1.2   Definitions of Cognitive Radio

While the focus for the cognitive radio technology on dynamic spectrum access is well justified and is acceptable within the wireless communication community, the original intent for this term was somewhat different. A typical expectation is that a cognitive radio is a smart radio whose capabilities include various aspects of artificial intelligence (AI), including perception of external inputs (not just communications signals), reasoning, decision support, self-adaptability, interoperability and such. Interestingly enough, the understanding of the term "cognitive radio" by an outsider to the wireless communication community is compatible with the original definition of this concept. In the following we provide a number of attempts to define cognitive radio.

The concept of cognitive radio was first introduced by Mitola in the late 1990s (Mitola and G. Q. Maguire 1999; Mitola 2000), and then refined to the following (as reported in Mannion 2006):

> A really smart radio that would be self-aware, RF-aware, user-aware, and that would include language technology and machine vision along with a lot of high-fidelity knowledge of the radio environment.

Clearly, this definition is asking for some AI technology in its implementation. Such a radio should be aware of itself and its environment (through sensors) and posses knowledge.

Since then, the definition of cognitive radio has been offered by a number of industry leaders, academia and others. Here are some of the CR definitions collected by the Cognitive Radio Working Group of the Wireless Innovation Forum (previously known as the Software Defined Radio Forum) (Cognitive Radio Working Group 2008).

The first of these definitions was adopted by the IEEE (DYSPAN P1900.1 Working Group 2008); it incorporates input provided by the Software Defined Radio Forum (Cognitive Radio Working Group 2007). This definition adds the decision making aspect to the previous definition and stresses the fact that such a radio would rely on predefined objectives.

a) A type of radio in which communication systems are aware of their environment and internal state and can make decisions about their radio operating behavior based on that information and predefined objectives.
b) Cognitive radio [as defined in item a)] that uses software-defined radio, adaptive radio, and other technologies to adjust automatically its behavior or operations to achieve desired objectives.

A similar definition has been accepted by the ITU Radio Communication Study Group (Radiocommunication Study Groups 2011). The above definitions have been refined by Haykin (2005).

Cognitive radio is an intelligent wireless communication system that is aware of its surrounding environment (i.e., outside world), and uses the methodology of understanding-by-building to learn from the environment and adapt its internal states to statistical variations in the incoming RF stimuli by making corresponding changes in certain operational parameters (e.g., transmit power, carrier frequency, and modulation strategy) in real-time, with two primary objectives in mind:

- Highly reliable communications whenever and wherever needed;
- Efficient utilization of the radio spectrum.

Software Defined Radio Forum has analyzed a number of definitions of the cognitive radio concept and published the summary of this study in Cognitive Radio Working Group (2008). Cognitive radio was considered from two points of view—design and implementation.

(Design paradigm) An approach to wireless engineering wherein the radio, radio network, or wireless system is endowed with *awareness, reason,* and *agency* to *intelligently* adapt operational aspects of the radio, radio network, or wireless system.

The terms *awareness, reason, agency* and *intelligently* are then explained as the capacities of collecting, classifying, organizing and retaining knowledge, applying logic and analysis of information (reasoning), and making decisions about operational aspects of the radio, network or wireless system so as to satisfy a purposeful goal.

1.3 From Definitions to Expected Capabilities of Cognitive Radio

Definitions are like high level specifications. How do they translate to the functional capabilities of radios? Here we first propose a summary of expected functional features of cognitive radio and then discuss some of them in more detail. This list extends the capabilities identified in Cognitive Radio Working Group (2008); the extensions come primarily from Kokar et al. (2009).

- *Sensing and information collection.* This implies that a CR should have access to information sources of possibly different modalities (e.g., GPS, vision, microphone, databases) and be able to integrate and represent this information into a coherent information structure useful to the radio.
- *Query by user or other radios.* Like an intelligent agent, a CR should be able to accept queries from its user and from other radios.
- *Situational awareness.* A CR should know not only the various parameters characterizing the environment, but should also be able to infer the existing relevant relationships and constraints between the different parameters and project the values and the relationships into the nearest future.

- *Self-awareness*. A CR should know its own parameters, variables and capabilities, and should be able to communicate this kind of information to other communication nodes.
- *Autonomous decision capability*. A CR should be able to understand goals and plan actions to satisfy or optimize the achievement of the goals while insuring that the actions do not violate any constraints imposed by either the design or by the external policies that apply to the locality in which the radio operates.
- *Query execution*. Like an intelligent agent, a CR should be able to reply to queries by providing the information it is asked for.
- *Command execution*. A CR should be able to interpret requests for, and to invoke specific behaviors (actions).

In the rest of this chapter we will discuss some of these features in more detail and then in the following chapters we provide a detailed explanation and examples of how some of them can be achieved. In particular, our focus is the implementation of constraint compliant optimization of the achievement and optimization of goals.

1.3.1   Sensing and Information Collection

Information can be collected by a CR either via sensing or via access to other information sources. Sensing in this context refers to the ability of a radio to collect information regarding its environment. Sensing can be performed locally and self contained in a radio or can be performed by other nodes located elsewhere in the network. Spectrum sensing, one of the most important sensing capabilities, is the process of measuring the characteristics of received signals and RF energy levels in order to determine whether a particular section of spectrum is occupied or not (Cognitive Radio Working Group 2008). However, in the current radio devices, the use of GPS is perhaps the most popular type of the sensing capability in use.

A CR can also access information available in other sources on the network. An example of such information is the information about the available frequencies stored in the white space database. According to the FCC's Memorandum Opinion and Order of September 23, 2010, wireless devices are allowed to use the white spaces (i.e., the spectrum allocated to the broadcasters but not used locally), provided the devices consult a white space database. Other examples of information available to a radio are regulatory policies that are applicable to the area that the radio is operating in, software components that a radio can download and install in order to be able to interoperate with other radios or security related sources that need to be accessed in order to verify the identity of a specific communication device.

1.3.2 Query by User or Other Radios

A cognitive radio may be aware of many things that the user is not. For instance, the radio may know how many other radios are operating within an area that it is in. Even more specifically, in a road traffic situation, it can know whether the other radios along the road are moving or not, and if moving, then at what velocity. Invoking its inference capabilities, it may derive the expected time to drive to a particular intersection. This kind of information may be useful to the driver for making a decision whether to continue on a specific road or take an exist and take a different road that under normal conditions would not be advisable.

Providing a feature for the radio that would constantly be monitoring such conditions may be not so wise because in some areas such congestion conditions do not happen so frequently. Thus a better solution might be to provide a capability to the radio user of formulating specific queries as needed—ad hoc queries. But in case a given query is of interest to the specific user very often, such a query could be added to the list of indicators to be monitored by the radio all the time.

Radios with the capability of query answering could also be queried by other radios. For instance, in the traffic situation described above, a radio could query another radio about what it already knows about the congestion of the traffic ahead of its location. While the querying radio may have its own assessment up to some point along the road, the radio that is ahead of it may complement its knowledge because it can reach further. Again, while such capabilities could be hard coded into the protocols, the query capability allows the radios to issue a practically infinite number of different types of query on demand, according to the specific needs of their users or other radios.

The capability of querying a radio may be provided in many different ways. In the most restricted case, a radio may have a number of queries hard-coded as functions into its software. Then either the radio user, or another radio, just call this function and provides the required input parameters. On the other hand, a much more flexible solution can be implemented in which queries are formulated in a query language. Examples of query languages are SQL (Structured Query Language) (Beaulieu 2009) for querying relational database systems and SPARQL (W3C 2006)—a standard query language for the Semantic Web. In this kind of solution, the query language allows the user or another cognitive radio to send a wide variety of queries expressible in the query language.

1.3.3 Awareness and Reasoning

"AWARE implies vigilance in observing or alertness in drawing inferences from what one experiences" (Merriam-Webster 2011). The translation of this definition to cognitive radio implies that a CR should be measuring (observing) the environment, inferring the consequences of what is being observed and be vigilant, i.e., react to the observations in terms of both a proactive collection of additional measurements and making and implementing decisions that are within the radio's control, e.g., modify its transmission parameters. For example, the radio might be monitoring which communication channels are free or occupied, the transmission energy levels and frequencies. Or it can monitor jammers and their characteristics.

Situation awareness is the next level of awareness in the sense that for a subject to be aware it must know not only about objects in the environment and their attributes, but also how these objects and attributes are related to other objects and attributes. Situation Awareness has been studied as a concept in Information Fusion. According to the definition most used in this domain (Endsley and Garland 2000), awareness is *the perception of the elements in the environment, the comprehension of their meaning, and the projection of their status in the near future.*

Situation awareness introduces an additional factor—the *goal*. The goal provides the focus to the observation process. Without a goal, an agent might observe every object that is possible to observe, limited only by the agent's abilities, and identify all possible relations among the objects. This would not be feasible and not necessary. Situation awareness thus is the identification of those objects and relations that are relevant to the agent's goal.

In the context of cognitive radio the radio would accept a query formulated by its user as its goal, e.g., "current traffic congestion along Route 128 between Route 93 and Route 90". Clearly, one possible solution for the radio would be to just query Google Maps for an answer to such a question. However, in case this is not available or insufficient, it also could identify other (mobile) radios in the area, infer their relation to this segment of Route 128 and then infer the answer to the congestion query for the user.

Awareness has been recognized as a feature of cognitive radio by the Wireless Information Forum (Cognitive Radio Working Group 2008), where "aware" was defined as that it "implies the ability to integrate sensations from the environment with one's immediate goals in order to guide behavior or draw conclusions." This report identified a number of types of awareness classified by the source of information: RF/Environment, Network, Location, User, Hardware, Policy and other. Additionally, the awareness of context (or situation) was also mentioned in the same report. Situation Awareness has also been listed as one of the most important features of cognitive radio in Kokar et al. (2009) where CR was considered as a cognitive agent.

1.3.4   Self-Awareness

Self-awareness refers to the ability of the radio to understand its own capabilities, i.e., to understand what it does and does not know, as well as the limits of its capabilities. In this way, the radio can determine whether a task is within its capabilities. In the case of a basic self-aware radio, it should know its current performance such as bit-error rate, signal-to-interference and noise ratio, multipath interference, etc. A more advanced CR would have the capability to reflect on its previous actions and their results. For example, a CR could be capable of analyzing the logs of past communication events and inferring the values of the various parameters that would guarantee better communication quality (Kokar et al. 2009).

It is important to understand that while any radio keeps the values of the various parameters (variables) in its computer memory, it does not mean that the radio actually "knows" about those parameters. For instance, even though the radio keeps the value of a variable "current SNR value", we cannot say that the radio is actually aware of the SNR value, similarly as we cannot ask a person whether the person has $100 in the pocket when the person is not aware of the contents of that pocket. Can we ask the radio "what is your current SNR?" and can we expect that the radio would answer such a question? Obviously, the radio's software could be designed in such a way so that the radio could accept such a query. But then the next question would be—what other queries would the radio be able to answer, such as the frequency offset, timing offset or equalizer taps? By examining such parameters, the radio could determine what parameter adjustments at the receiver could improve the performance of the communication link. Then two communicating radios could negotiate how to adjust these parameters in order to better achieve their communication goals.

The self-awareness of own capabilities of the radio can be achieved by using what is called "reflection", for instance, Java reflection (Moskal et al. 2010; Moskal 2011; Wang et al. 2004). Java reflection provides a means to query the internal structure and the values of the internal variables of a Java executable. For instance, a running program can query itself about the Java classes that constitute the program, the attributes of the classes and the methods that a particular class implements.

1.3.5   Autonomous Decision Capability

Being *autonomous* means "having the right or power of self-government; undertaken or carried on without outside control; self-contained; existing or capable of existing independently; responding, reacting, or developing independently of the whole" (Merriam-Webster 2011). Autonomy is the distinguishing feature of *agent*, as compared to other software items, like program, or module. In artificial intelligence, agent refers to an autonomous entity which observes and acts upon an environment and directs its activity towards achieving its own goals (Russell and Norvig 2003). Here we interpret "its own goals" as the subgoals that are decomposition of a higher level goal programmed into the agent.

In the context of cognitive radio, the goal might be "maximize throughput" measured, for instance, in terms of the application layer data bitrate. Other goals may be "avoid detection" or "avoid incumbent interference" (cf. Li and Kokar 2010). The latter goal refers to the opportunistic use of spectrum by unllicensed radios. Since the spectrum is owned by specific service providers, the radios operating under a license from the providers are referred to as "incumbents". The unlicensed radios should not cause interference to the incumbents. The non-interference goal may be hard-coded into a standard software defined radio. However, a cognitive radio could have such a goal provided to it via a policy, rather than hard-coded. In order to satisfy such a goal, the CR would need to have the capability of interpreting policies and executing the required protocols defined by the policies.

1.3.6 Query Execution

The query execution capability is directly dependent on the type and the flexibility of the querying mechanism supported by a CR. In the case when a radio supports only a limited number of pre-defined queries, each of the queries can be implemented as a function that takes the query parameters and returns the results. However, in the case of supporting a query language, the CR must be able to parse the query, search its own data, find matches to the query and return the data to the querier. If the query is about data related to the environment, this data must be available in the radio's data base. This in turn requires that the radio has some capability of logging data that it collects during its operation. If the query is about the radio itself, the radio must be able to not only understand which data it is asked about, but also where it can found in its own data structures. To achieve this, the radio must have a representation of its own model and be able to access its own variables, which in turn implies that the radio must have the reflection capability as discussed in Sect. 1.3.4.

1.3.7 Command Execution

Similarly to queries, commands can be implemented either as, for instance, a list with associated functions that execute the particular commands or as a capability to accept any command expressible in a formally defined command expression language. For cognitive radio the latter approach is more appropriate since a cognitive radio, as discussed earlier, must be able to deal with the circumstances that were not anticipated at the design time.

In the context of cognitive radio, command execution means the capability of invoking functions encoded in the radio software. In the most extreme case this would mean that each subset of functions that a radio supports would be invokable on demand and in any sequence. Such a solution is probably not feasible, since it would interfere with the execution of the functionality that implements a particular waveform. A less aggressive solution would allow CRs to modify their internal parameters on demand by either the radio user or other radios. A technical solution to implement such an approach was discussed in Moskal et al. (2010). An example of a radio parameter that can be modified on demand, the radio transmit power, was discussed in Li et al. (2011).

1.4 Autonomous Adaptation/Optimization

Having introduced the main concepts we are now ready to discuss the main subject of this book—the flexible, collaborative adaptation in the cognitive radio. The implementation of flexible adaptation requires a combination of a number of features discussed above.

In the context of cognitive radio, adaptation means adjusting operational parameters of a radio in order to optimize its performance in terms of a specific measure. The operational parameters are changed according to an algorithm based on the feedback that is either directly provided to the radio by the environment or calculated as a function of a number of measurements and values of the radio's internal parameters. An example of the performance measure can be bit error rate (BER). An example of the operational parameter is transmit power.

Adaptation is a special case of feedback based control, as defined in control theory (cf. Åström 1989). In a typical use of control theory, there is a *plant* whose parameters are manipulated by a *controller*, based on the feedback received from the plant in order to achieve a control *goal*. In communications, the situation is usually more complicated since the controller in any radio needs to make its control decisions not only on the feedback from the environment, but also on the information from other communication nodes. For instance, a transmitting node needs to have some information from the receiving node on the quality of the received signal, e.g, by monitoring the frequency of retransmit requests.

Collaborative adaptation is built into every communications protocol. For instance, radios adjust the length of the equalizer in order to keep the value of the equalizer error (and thus the BER) at an acceptable level. Communication protocols are structured in such a way that information needed for adaptation is encoded into the specific fields of the packets.

The control goal in such adaptations is also built into the radio's software. It is expressed in terms of a performance measure. For example, the goal may be to minimize the BER. The next level of flexibility is to give the radio the capability of dynamically changing control goals. As mentioned above, this operation would also have to be synchronized with other nodes involved. To achieve this objective, radios would need the capability of negotiating goals and transmission parameters. This capability is the main focus of this book.

1.5 Organization of This Book

In Chap. 2 we will describe the conceptual architecture of cognitive radio, then introduce the concept of Ontology-Based Radio (OBR).

In Chap. 3, we will first classify cognitive radio as an agent and then introduce how cognitive radio agent observes the environment and adapts its behavior accordingly. Then we will discuss the role of collaboration in cognitive radio adaptation and address the necessity of collaborative adaptation.

In Chap. 4, we will discuss the signaling options to implement collaborative adaptation and conclude that ontology-based signaling plan enables the cognitive agents to exchange and interpret knowledge and execute high-level missions. We also propose the metrics to evaluate the flexibility of a signaling plan. By comparing the selected signaling options using these metrics, we conclude that the ontology-based signaling brings a great flexibility to the communications.

In Chap. 5, we will give a brief introduction of FIPA Agent Communication Language (FIPA ACL), along with a detailed description of the ACL message structure and the communicative act library.

In Chap. 6, we will describe a link adaptation use case that will be used to demonstrate how to do collaborative radio adaption to improve link performance.

In Chap. 7, we will first discuss the concept of knowledge and inference, then start to compare the knowledge-less approach and knowledge-rich approach. After that, we will discuss the language selection issues for ontology and policy. In order to standardize the ontology-based approach to the cognitive radio communities, we participated in the standardizing work in the Wireless Innovation Forum and developed a Cognitive Radio Ontology (CRO) to capture the basic terms of wireless communications. The CRO has been approved by the Wireless Innovation Forum as its recommendation and it is expected to provide a standardized way of representing signaling among cognitive radio.

In Chap. 8, we will outline the top-level structure of the CRO and describe the design principles and methodologies we used in the process of ontology development.

In Chap. 9, we will focus on the link adaptation use case and describe our MATLAB simulation and GNU/USRP implementations of the Ontology-Based Radio for this use case. In Chap. 10, we will use this use case to evaluate the benefits and costs of the ontology-based signaling.

Chapter 2
Cognitive Radio Architecture

In this chapter we discuss some of the architectural constraints and boundaries within which a cognitive radio operates. We begin by listing the interfaces that a cognitive radio uses or implements. Then we present a brief overview of the cognitive architectures studied (primarily) in Artificial Intelligence. We then present an example architecture developed by the DARPA XG program and used as a reference in the IEEE P1900.5 standardization effort. Finally, we briefly discuss software defined radio as a platform for the implementation of cognitive radio.

2.1 Cognitive Radio Interfaces

A typical nowadays radio interacts with a number of external systems, including the radio user, the network (via a base station), sensors (e.g., spectrum sensor, GPS) and other resources accessible through the network, e.g., other users (social network), informational resources, like the Web resources, or more generally the Information Cloud. All these interfaces are shown in Fig. 2.1. Most of these interfaces were recognized as features of cognitive radio in Fette (2009).

Intelligent interaction with the user. In addition to the usual support for the transmission of audio signals, cognitive radio may support voice recognition and perception, e.g., recognition of queries and commands. This capability is already available in today's smartphone technology. Going farther, it could use speech recognition technology to perceive conversations, retrieve and analyze the content of conversations and give advice to the user (Mitola 2009b).

Interaction with sensors. The next step is vision and image processing. For example, a cognitive radio could use vision algorithms to understand the world around the user and detect opportunities to assist the user using this information.

Currently, the allocation and utilization of spectrum follows a "command and control" structure which is dominated by long planning cycles, assumptions of

S. Li and M.M. Kokar, *Flexible Adaptation in Cognitive Radios*,
Analog Circuits and Signal Processing, DOI 10.1007/978-1-4614-0968-7_2,
© Springer Science+Business Media, LLC 2013

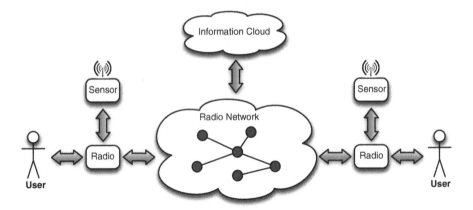

Fig. 2.1 External interfaces of cognitive radio

exclusive use, conservative worst case analysis and litigious regulatory proceedings. Using spectrum-aware radios, the management of spectrum could be transitioned into a new structure that is embedded within each individual radio. Collectively, implicitly or explicitly, the radios would cooperate to optimize the allocation of the spectrum to meet RF devices' needs (Marshall 2009). To achieve this objectives, radios would have to have the spectrum sensing capability. This could be realized by a spectrum sensor that is part of the radio.

Other sensors could also be installed on the radio. In nowadays' radios, the GPS sensor is typically used to support such applications as localization of the user's radio, mapping and direction finding applications.

Intelligent interaction with the network. Cognitive radio could provide standardized interfaces to access heterogeneous networks and support the management and optimization of network resources (Smith 2009). Referring to the OSI protocol stack, a cognitive radio's capabilities can be allocated to the particular layers:

- Application layer: It is an advanced PDA (personal digital assistant) with communications capability. As such, the radio would have to support simple, yet rich in functionality, interfaces for the user.
- Network layer: It should allow for an easy interaction with the heterogeneous networks that the radio can reach.
- Data link and physical layers: Cognitive radios could provide better performance in terms of connectivity, transmission capacity, reach, bandwidth, spectrum and other aspects of communication.

In this book we are focusing on the capabilities of cognitive radio that can be attributed to the Data Link and Physical layers.

Interaction with other resources on the network. Other resources available to the radio on the networks include, among others, two types of virtual networks—social networks and information networks. Social networks are groups of users who are related via the various social relationships, e.g., *knows*, *trusts*, *manages*, and many others. The *knows* relationship is a collection of links between pairs of humans, such that one of them knows the other. Similarly, the *trusts* relationship captures the links between persons who trust other persons. Such relations can be viewed as *links*, although not as the links in the communications networks. For such links to be materialized in the communications networks, the human users must be connected to the network and the nodes to which they are connected should be linked by an active communications link.

Similarly, an *information network* can be realized as a virtual network over a communications network. The nodes in such networks are information stores, e.g., databases. The links in such networks are relations among the particular data stores, or even data items in the stores. For instance, one database may contain information about houses in a given neighborhood. Another database may contain information about people. And yet another database may contain information about the financial liabilities of particular people. These three databases are related via some relationships. For example, the people database is related to the houses database via the address where the person lives, while the liabilities database is related to the people database via the social security key. In effect, the liabilities of the persons can be traced to the financial liabilities against houses.

In today's computation and communication environments, a lot of data is stored "in the cloud", i.e., on the servers that the user even does not necessarily knows where they are. However, many users want to have a more direct control over their information resources. This issue is related to the security of information exchange. Cognitive radio might need to provide interfaces to both the user and to the cloud so that the user could have control over the whereabouts of its information resources.

2.2 Cognitive Architectures and Processes

Since cognitive radio has "cognitive" in its name, it is natural to relate the notion of cognitive radio to the cognitive architecture and processing. A cognitive architecture specifies a structure of an intelligent system (or agent). Usually, it also specifies a processing sequence for such agents. Research has been done on the topics of cognitive architecture and processing in a number of communities—cognitive science (Laird et al. 1987; Anderson et al. 2004), human factors (Endsley and Garland 2000), artificial intelligence (Langley et al. 2009), intelligent control (Albus et al. 1992; Taylor and Sayda 2005), cognitive networks (Clark et al. 2003; Mahmud 2007) and cognitive radios (Mitola 2009a). The objectives of this research were somewhat different. While the focus for cognitive science is to develop a computational model for human cognition, sometimes referred to as *the mind*, the goal for AI is mainly to build machines that just exhibit the "intelligent" behavior. For cognitive networks and cognitive radios, this goal is even more specialized— to exhibit intelligent behaviors in the technical devices, like routers, handsets and basestations.

Cognitive architectures identify blocks, like short-term and long-term memories, possibly organized into some more complex representational structures, as well as functional processing blocks. Memories contain information collected via sensing as well as agent's beliefs, goals and knowledge. Processing blocks include functions for manipulating acquired information, learning new knowledge, modifying beliefs, controlling collection of information and so on. Below, we briefly overview some of the examples of cognitive architectures known in the literature.

ACT-R (Anderson et al. 2004) is an architecture that consists of processing modules, each specialized to process a different kind of information—sensory information, beliefs, goals, actions, declarative knowledge. Associated with modules are buffers, which serve as short-term memories. Information is stored in structures called "declarative chunks". The long-term memory contains *production rules* consisting of *condition* parts and *action* parts. Chunks include some information on their past usage; rules contain some information on their utility. Actions are executed when conditions are satisfied. Actions include both modifications of the current information structures as well as execution of some external actions, e.g., motor commands. ACT works in cycles. In each cycle it matches rules to the contents of the short-term memory and selects and executes production rules, which is guided by the goals and by the utility of the productions. The learning process can modify the production rules, both parametrically and structurally. ACT-R has been used extensively to model various aspects of human behavior.

SOAR (Laird et al. 1987) was one of the first architectures developed in the cognitive science and AI communities. The philosophy of SOAR is based on eleven hypotheses, including, among others, that intelligence must be realized with a symbol system, control is (symbolic) goal oriented, all declarative knowledge is stored in a uniform representation, long-term knowledge can be encoded by a production system, knowledge is organized (and learned) in chunks associated with the goal structures, goals can be created dynamically, weak methods (i.e., methods that arise from the interaction with the task, rather than being programmed) are part of the architecture.

The main components of the SOAR architecture are Problem Spaces (sets of states and operators that manipulate states), Long-term Production Memory (condition–action rules), Short-term Memory (context stack—objects with attributes and their values) and Preference Memory (preferences for particular elements of the context stack). SOAR's execution cycle includes two phases: elaboration (execution of all the matching productions) and decision (analysis of the preferences and putting of next problem space, operator or goal in the context stack). The cycle is repeated until the goal is reached.

Since there are so many proposed cognitive architectures, an attempt was made to develop a *schema* for instantiating particular cognitive architectures. The schema was called *CogAff* (for Cog-nition and Aff-ect) (Sloman 2001). *H-Cogaff* was a specific instance of CogAff. This architecture is presented as a grid—three columns and three layers. The layers correspond to Reactive mechanisms, Deliberative reasoning and Meta-management (or reflective processes). The columns correspond to the Perception–Central Processing–Action information processing cycle.

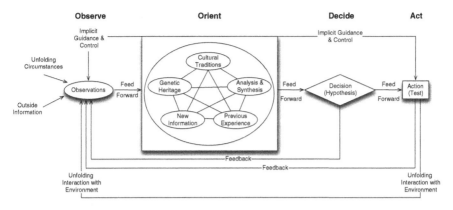

Fig. 2.2 The OODA loop

The processing cycle of CogAff abstracts what is part of many of the cognitive architectures—the three main processing steps: perception, cognition and actuation. The same pattern was advocated for intelligent controllers (cf. Antsaklis and Passino 1992), where the three processing steps were called the *perception, reasoning, action triad*. A similar idea was advocated by Boyd as a way of processing information with the goal of winning in military situations. Boyd proposed what is known as the OODA loop (Boyd 1987). The OODA loop (see Fig. 2.2) contains the four major steps: Observe, Orient, Decide and Act. Each of the steps may include other loops. Moreover, it is not a simple loop since the particular steps interact by passing both data and control information.

The Observe and Act steps loosely correspond to the Perception and Action steps in the various cognitive architectures. The mapping of the Orient and Decide steps to particular architectures would need to be analyzed on the case by case basis. It is worth noting that the Orient step includes many functions, primarily knowledge-based processing.

The OODA loop was embraced by the Information Fusion community, where it serves as a basis for situational awareness (cf. Endsley and Garland 2000). Endsley's model of cognitive situation processing includes three steps: Perception, Comprehension and Projection, where Perception corresponds to the Observe step in OODA. Comprehension and Projection correspond to Orient. Projection is the step that derives a projection of a situation into the nearest future.

2.3 Cognitive Architectures and Cognitive Radios

The use of the OODA loop for cognitive radio has been proposed by Mitola and G. Q. Maguire (1999). These ideas have been mentioned in many publications. Recently, Moy (2010) used the OODA loop to pattern the design cognitive radio upon it.

In Amanna and Reed (2010) the various cognitive architectures were reviewed (e.g., SOAR, ACT-R) as well as architectures of cognitive radios [e.g., CBR (He et al. 2009), Public Safety Cognitive Radio (Le et al. 2007), Open Source Cognitive Radio (OSCR) (Stuntebeck et al. 2006), DARPA xG]. However, only two of the cognitive radio solutions (architectures) were classified as being related to the OODA loop and one solution to the SOAR architecture. The general conclusion was that the developments in the area of cognitive radios is not sufficiently integrated with the developments in the domain of cognitive architectures.

The architecture of the radio developed under the DARPA xG program (cf. Perich et al. 2010), was also patterned upon the OODA loop. The cognitive capabilities of the xG radio, similarly as the most of the other cognitive radios, was limited to the functionality for achieving dynamic spectrum access.

2.4 Ontology Based Cognitive Radio

In our work we use the Ontology Based Radio (OBR) paradigm for the architecture and development of cognitive radio, introduced in 2003 (Wang et al. 2003). The main idea of OBR is flexible signaling, i.e., control messages are embedded in the payload of communication packets and the messages are expressed in terms of an ontology. In this way, control message types are not limited to the specific protocol, but instead, can be of any type expressible in the language defined by a given ontology. Unlike other radios that utilize ontlogies, e.g., Perich et al. (2010), OBRs are not limited to a specific Application Programming Interface (API), but instead, are based on an API bound to a specific language, e.g., Web Ontology Language (OWL) (Schreiber and Dean 2004). Also, OBRs are self-aware due to the use of the feature of *reflection*, e.g., Java reflection (Wang et al. 2004). The OBR approach will be described in more detail later in this book.

The internal processing in OBR is based on the OODA loop. The "observe" part of the loop includes not only spectrum measurements, but also inputs from other sensors and from other sources of information, as shown in Fig. 2.1. Moreover, OBRs can actively seek information by sending queries for specific types of information to other radios and even requesting other nodes to collect some information on demand.

The behaviors of OBRs, like most ontology-based radios, is controlled by *policies*, i.e., collections of *event–condition–action* rules. Radios monitor *events* and react to the events according to the *actions* specified in the policies that match the applicability *conditions*.

Since ontologies allow for expressing very complex types of messages and since policies allow for expressing very complex behaviors, the ontologies and the policies may need to be structured and modularized in order to make the OBR approach practical and scalable. Towards this aim, a number of architectural solutions have been proposed. One of the approaches (cf. Raymer et al. 2006), was to have a *continuum* of policies, with each component policy being crafted to a particular constituency in the communications chain. In the case when the ontologies and policies use different languages, this approach would require a continuum of inference engines, one for each ontology/policy language.

While the OBR approach does not preclude such a compartmentalization of policies, it does not, however, provide any special architectural features to support such a fine partitioning of policies. Instead, this kind of partitioning would need to be resolved with the means of a particular ontology and policy representation language. For instance, OWL supports the *import* feature, i.e., particular ontologies can be added to a given ontology via this operation.

The ontology-based radio solutions developed under the DARPA xG program (cf. Perich et al. 2010) and the reference architecture included in the IEEE P1900.5 standard (Group 2011) adopted the approach in which two separate inference engines are identified. This architecture was influenced by Marshall's partitioning of the policies and inference modules into *endogenous components* and *exogenous components* (Marshall 2009). An exogenous component executes and enforces external policies. It addresses the radio's impact on the external environment, and ensures that the behaviors of the radio satisfy the constraints imposed by external regulations and policies. For example, an exogenous component can assist the radio in avoiding spectrum interference while searching for spectrum opportunities. Conversely, an endogenous component internally optimizes the performance of the radio through selection of operating mode and other parameters.

Based on the above perspective, the basic architecture of a cognitive radio that addresses the distinction between endogenous and exogenous components can be viewed as in Fig. 2.3. The abstract architecture of a cognitive radio comprises eight components (Denker et al. 2009; Stewart 2009; IEEE P1900.5 Working Group 2011):

1. Sensors. In a cognitive radio that implements DSA, sensors are used to collect information from the external environment and discover available spectrum and transmission opportunities.
2. Radio Frequency (RF). The RF component is used to transmit and receive signals.
3. Radio Platform. The radio platform includes the digital signal processing and the software control. It provides interfaces to communicate with the RF, sensors, information source and sink, and the policy reasoners.
4. System Strategy Reasoner (SSR). The SSR is an *endogenous component* of the cognitive radio. It forms strategies to control the operation of the radio. The strategies reflect the spectral opportunities, the capabilities of the radio and waveform, and the needs of the network and the users.
5. Policy Conformance Reasoner (PCR). The PCR is the *exogenous component* of the cognitive radio. It executes the active policy set to ensure that the radio transmission conforms to the policy.
6. Policy Enforcer (PE). The PE acts as a gate keeper between the SSR and the Radio Platform. It ensures that all the transmission decisions sent from SSR to the Radio Platform comply with the active policy.

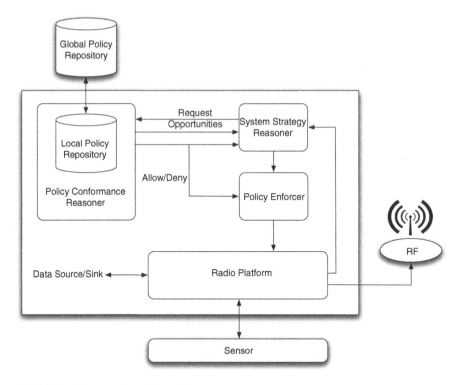

Fig. 2.3 Architecture of cognitive radio

7. Global Policy Repository. The Global Policy Repository stores all the policies
   and specific subsets configured for specific networks. The Global Policy Repository is shared across the network.
8. Local Policy Repository. The Local Policy Repository is within the SSR. It can
   download the policies from the Global Policy Repository through an interface.
   A radio node can store multiple sets of policies, but only one set of policies is
   active at any time.

In this book we focus on the functionality and the use of the SSR since this
is the component that is actively involved in the process of adaptation of radio's
communication parameters. Below we describe the interactions of the SSR with the
other architectural components of a cognitive radio that is implemented according
to the architecture shown in Fig. 2.3.

As was mentioned earlier, the OBR repetitively executes the OODA loop. The
packages that carry either voice or data are processed by the Radio Platform.
Symbolically, in Fig. 2.3, this is shown as the path to the Data/Source Sink. Our
real interest is in the control messages—the signaling. Those messages are extracted
by the Radio Platform and sent to the System Strategy Reasoner (SSR). The SSR
executes the "Orient" and "Decide" steps of the OODA loop. When the SSR infers

that a packet needs to be sent to the Radio Platform (and then to the RF component), it may need to send a query to the Policy Conformance Reasoner PCR) and ask whether the transmission is allowed or not.

The interaction between the SSR and the PCR is implementation dependent. Here we describe how this kind of interaction was specified in Stewart (2009) and Denker et al. (2009). According to these descriptions, PCR's decisions are called *transmission opportunities* that are valid for a period of time. The SSR sends a query/request to the PCR when the radio needs to change its transmission strategy, e.g., after the validity time period for a permitted transmission opportunity expires. The SSR may send one of at least two types of transmission request.

* *Unbounded transmission request.* The SSR asks the PCR to assist in identifying transmission parameters that are policy compliant. The request may not have values specified for all transmission parameters. For example, the SSR may ask: "I want to send a packet to Radio_B at time_T and at place_P—which waveform should I use?" The PCR identifies the transmission parameters that meet both the needs of the SSR's request and comply with the active policy set. Then, the PCR sends a reply back to the SSE. The reply includes the transmission parameters such as transmission power, frequency, data rate, modulation, and so on.
* *Bounded transmission request.* The SSR sends a fully bounded transmission request to the PCR, i.e., all the values of the transmission parameters are explicitly specified. The PCR evaluates the request to assess whether it complies with the active policy set and passes the result to both the Policy Enforcer (PE) and the SSR. The result can one of three types: (1) the transmission request is allowed; (2) the transmission is not allowed; (3) the transmission is allowed if specified additional constraints are added. The constraints may be acceptable values of the underspecified request parameters.

All the outgoing control messages generated by the SSR are passed to the Policy Enforcer which then ensures that all the transmissions conform to the policy. The Policy Enforcer forwards the control message to the Radio Platform. The outgoing data message and control message are merged in the Radio Platform, and then sent out through the RF.

The SSR can also send control messages to the Sensor. The sensor then collects the information of the environment and discovers which parts of the spectrum are occupied or not, accordingly.

2.5 Cognitive Radio Platform: Software-Defined Radio

In this section we provide a brief overview of software defined radio (SDR)—the platform for the implementation of cognitive radio. The definition of SDR is given by IEEE SCC 41-P1900.1 as the "radio in which some or all the physical layer functions are software defined" (DYSPAN P1900.1 Working Group 2008). The properties defined by software include carrier frequency, signal bandwidth, modulation, network access, cryptography, channel coding (e.g., forward error correction coding) and source coding (voice, video and data). SDR is a general-purpose platform that can adapt to a wide range of waveforms, applications and products. Different kinds of waveforms at different frequencies can be implemented on the same SDR processor. Thus SDR is cost effective, versatile and easy to upgrade (reduced development cycle time) (Fette 2009).

Typically, an SDR consists of a stack of hardware and software functions, each with open standard interfaces. The SDR hardware architecture usually consists of the RF Front End, A/D converter, and the Digital Back End. First, the RF Front End amplifies the received signal, and then converts the carrier frequency of the signal to a low intermediate frequency. Second, the A/D converter converts the analog signal to a digital signal proportional to the magnitude of the analog signal. Third, the digital signal is further processed by a digital signal processor (in the Digital Back End) to perform the modem (modulation-demodulation) functions (Fette 2009).

The RF Front End usually consists of receiver and transmitter analog functions such as frequency up-converters and down-converters, filters, and amplifiers. In the full-duplex mode, the RF Front End provides some filtering to prevent the high-power transmitted signal from interfering with the lower-power received signal (Robert 2009).

The Digital Back End consists of General-Purpose Processors (GPP), Field-Programmable Gate Arrays (FPGAs) and Digital Signal Processors (DSP). A GPP usually performs the user applications and high-level communications protocols, whereas a DSP is more efficient in terms of signal processing but is less capable to process high-level communications protocols. The FPGA complements DSPs in that it provides timing logic to synthesize clocks, baud rate, chip rate, time slot and frame timing, resulting in a more compact waveform implementation. In general, the SDR hardware design is a mixture of GPPs, FPGAs and DSPs to provide a flexible platform to implement various waveforms and applications. Dedicated-purpose Application-Specific Integrated Circuits (ASIC) are not particularly suitable for SDR hardware due to their lack of flexibility (Fette 2009).

The Digital Back End is used to implement functions such as modem, Forward Error Correction (FEC), Medium Access Control (MAC) and user applications. The modem converts symbols to bits by a sequence of operations. First, the digital down-converter (DDC) converts the digitized real signal centered at an intermediate frequency to a baseband complex signal at a lower sampling rate. Second, the signal is filtered to the desired bandwidth. Next, the signal is time-aligned, despreaded and re-filtered. Then, a symbol detector is used to time-align signal to symbols. An equalizer is also used to correct for channel multipath effect and filter delay distortions. Finally, the symbols are mapped to bits using the modulation alphabet. Due to interference, the signal may be received with errors. FEC uses the redundancy introduced in the channel coding process to detect and correct the errors. FEC can be integrated with the demodulator or the MAC processing. After the MAC layer processing and network layer processing, the data are passed to the application layer that performs user functions and interfaces such as speaker/microphone, GUI, and other human-computer interfaces. The user application layer usually includes vocoder, video coder, data coder and web browser functions. Typically, voice applications are implemented in DSP. Video applications are usually implemented on special-purpose processors due to the extensive cross-correlation required to calculate the motion vectors of the video image objects. Text and web browsing usually run on GPP (Robert 2009).

On top of the hardware, several layers of software are installed, including the operating system, boot loader, board support package and the Hardware Abstraction Layer (HAL). It is essential to present a set of highly standardized interfaces between the hardware platform and the software, and between the software modules so that the waveform and applications can be installed, used and replaced flexibly to achieve the user's goals (Fette 2009).

There are two open SDR architectures—Software Communication Architecture (SCA) and GNU radio (Robert 2009). SCA is a standardized software architecture sponsored by the Joint Program Office (JPO) of the US Department of Defense (DoD) for secure signal-processing applications on heterogeneous, distributed hardware. It is a core framework to provide the infrastructure to create, install and manage various waveforms, as well as to control and manage the hardware. In addition, it provides a set of standardized interfaces to enable the interaction with external services (US Joint Program Executive Office 2011; Snyder et al. 2011; Communications Research Centre Canada 2011).

GNU radio is a Python-based architecture that provides a collection of signal processing components to build and deploy SDR systems. It is designed to run on general-purpose computers on the Linux operating system (Blossom 2004).

Chapter 3
Collaborative Adaptation

3.1 Cognitive Radio Agent

As was stated in Sect. 1.3, cognitive radio can be viewed as a autonomous agent. An agent is an entity that perceives its environment through sensors and acts upon that environment through actuators in order to satisfy its goals (Russell and Norvig 2003) (see Fig. 3.1).

For instance, a taxi driver agent perceives the road environment through sensors such as the cameras, speedometer, GPS, or microphone. Based on the information collected from the sensors, the driver then maps the perception to a sequences of actions. The available actions include controlling the engine through the gas pedal and controlling the car via steering and braking. The mapping from the perception to the actions specifies which action an agent ought to take in response to a given perception. For example, the driver agent ought to brake when it perceives a red light. This mapping describes the behavior of the agent. However, in some of the cases, knowing the current state of the environment is not suffficient to decide which action to take. For example, the taxi can turn left or right at a road junction, depending on to which destination the taxi is going. That is, besides the current state of the environment, some goal information must be provided to the agent in order to make the decision. The goal information describes the desirable state, such as the passenger's destination. Once the goal changes, the actions may change accordingly.

3.2 Knobs and Meters

The interactions between the agent and the environment shown in Fig. 3.1 must be supported by the agent's interfaces. For the radio domain, those interfaces are known as *knobs* and *meters* (Rondeau 2007). Meters refer to the variables in the radio software whose values represent the feedback received from the environment. The knobs, on the other hand, are the variables in the radio software that determine the characteristics of the signal emitted by the radio.

S. Li and M.M. Kokar, *Flexible Adaptation in Cognitive Radios*,
Analog Circuits and Signal Processing, DOI 10.1007/978-1-4614-0968-7_3,
© Springer Science+Business Media, LLC 2013

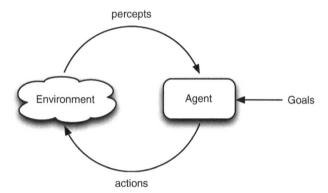

Fig. 3.1 Agent interacts with the environment

In other words, we can think of the radio as having adjustable knobs that control the radio's operation and thereby affect the radio performance. Meters refer to the utility or cost functions that are intended to be maximized or minimized in order to achieve optimum radio operation.

The performance, or the *QoS*, of the radio can be calculated based on the meter readings. The way to assess the QoS varies depending on the application. For example, based on the same meter reading, the calculation of QoS is different for voice communication, web browsing, or video conference.

Knobs and meters in the radio have complicated dependency relationships, i.e. knobs affect certain meters in different ways. For example, increasing the order of the modulation scheme increases the data rate, but decreases the BER. In Rondeau (2007), Rondeau provides a detailed analysis of the dependency relationship between different meters.

A list of typical knobs and meters in cognitive radio can be found in Cognitive Radio Working Group (2008).

3.3 Feedback-Control Model for Adaptation

In Sect. 2.4 we introduced the OBR as an OODA loop based system. In engineering, the OODA loop is often implemented as a feedback control loop. While in engineering those systems involve hardware, a similar approach has been proposed in Kokar et al. (1999) for software systems.

Figure 3.2 shows the closed-loop feedback control model for the radio. In this model, the system is split into a controller, a plant and a QoS subsystem. Recalling the cognitive radio architecture in Fig. 2.3, we can think of the SSR as a controller. The controller calculates the desired values for the knobs as a function of the goal and the QoS. Then, the plant, being the actual operational part of the radio, uses the values of the knobs in the transmission actions. Note that the actions of the plant

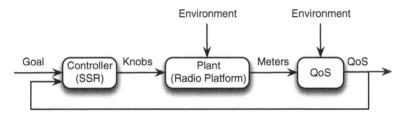

Fig. 3.2 Example of cognitive radio control model

are subject to the disturbances, represented in Fig. 3.2 inputs from Environment. Some other inputs from Environment are received through the radio receiver and thus are reflected in the values of the meters. These meters are used by the QoS subsystem for computing the value of the QoS that reflects the overall performance of the system, which then is used as feedback. Based on the goal of the application and the QoS, the controller computes the value of the input (knobs) to the plant to achieve the control goal.

3.3.1 Adaptation Levels in Feedback Controller

There are different levels of adaptation in the feedback controller shown in Fig. 3.2 (Polson 2009).

1. At a low level, the adaptation algorithm is built into hardware. For instance, in 802.11a, radios are able to sense the bit error rate and then adapt the modulation to a data rate and then forward error correction (FEC) such that the bit error rate can be controlled at an acceptable low level. This algorithm is implemented in application-specific integrated circuits (ASIC) chips.
2. At an intermediate level, the adaptation is software-defined. One way to achieve it is to hard code the adaptation algorithm into the radio. The shortcoming of this approach is that the algorithm is hard-coded into the radio and forms an inseparable part of the radio's firmware. Another way is to write the adaptation algorithm into a set of policies that control the radio behavior. This approach separates the adaptation policies from the implementation and thus exhibits more flexibility on the modification of the adaptation algorithm.
3. At a high level, the radio is able to learn from its experience and adapt its parameters without human interventions. Learning means that when the system is presented with a set of environmental test stimuli, the decisions it arrives at are not constant, but improve with time and experience. A typical example of learning is the case-based reasoning. The radio records the perception, the action and the result of each case from its past experience. In this way, the radio will gradually learn more about the environment, and better adapt to the

environment. By learning we mean that if presented with the same set of input conditions, a radio with learning capability may react differently depending on how it perceives the environment, whereas a radio without learning capability will always arrive at the same conclusion regarding how the radio should operate (Cognitive Radio Working Group 2008).

3.4 Role of Collaboration in Cognitive Radio Adaptation

As was mentioned in Sect. 1.4, in communications, the controller in any radio needs to make its control decisions not only on the feedback from the environment received either via its sensors or the radio's receiver subsystem, but also on the information from other communication nodes. For example, a transmitting node needs to have some information from the receiving node on the quality of the received signal. This kind of information cannot be simply evaluated based on the signal that the radio receives since it depends on the quality of the signal sent by itself and received by the other radio. Some of this capability is implemented in the communications protocols. For example, radios adjust the length of the equalizer in order to keep the value of the BER at an acceptable level.

Another piece of information that cannot be estimated by a radio locally is the control goal of the other radio. If the goal is built into the protocol, e.g., minimize the BER, and the designer of the software radio knows the goal, the designer then implements the software based upon this knowledge. The next level of flexibility would be to allow cognitive radios change their goals during their operation. However, the operation of dynamically changing goals would have to be synchronized with other nodes involved. To achieve this objective, radios would need the capability of negotiating goals and transmission parameters. In other words, radios need to collaborate.

In more general terms, collaboration is the process in which an agent tries to achieve a common goal shared by a community and ensure the community acts in a coherent manner by reasoning about its local actions and the (anticipated) actions of others (Jennings 1996). As discussed in Xuan et al. (2001), agents must communicate with other agents to obtain the non-local information in order to make decisions of what actions to take. Collaborative adaptation helps an agent to obtain a complete view of the environment and better achieve the adaptation goal.

The objectives of collaborations are listed in Durfee et al. (1990):

- Forming a solution faster by working parallely
- Improving problem solving by allowing agents to exchange information
- Reducing communication costs by exchanging more selective messages
- Avoiding duplicate efforts by letting the agents to recognize redundant activities

Among the above objectives, improving problem solving by exchanging information between agents is of our particular interest. In cognitive radio systems, there are a few reasons why collaboration is required. In Jennings (1996), the author summarizes the reasons why multiple agents need to coordinate their actions. Since cognitive radio system can be viewed as a multi-agent system, Jennings' analysis can also apply to cognitive radio systems. Below we list the reasons why cognitive radio systems need to be collaborated:

1. There are complicated dependencies among the knobs and meters in cognitive radio, which leads to a situation that the goals undertaken by the individual radios are related. In some cases, the decisions made by one radio will have an impact on the decisions of other radios. For instance, the decision on the length of the training sequence depends on the length of feedback/feedforward taps at the equalizer. If the training sequence is too short, then the equalizer does not fully converge and thus CRC errors may appear. Conversely, if the training sequence is too long, it will have little effect on equalizer performance.

    In other cases, two radios may act as two hostile agents, trying their best to achieve the same goal. For instance, two radios are in the same environment and trying to maximize their throughput. Each radio can implement a narrowband waveform or a wideband waveform. If both of them implement the wideband waveform, then there will be more interference and none of them is able to achieve its own maximum throughput (Neel et al. 2009).

2. There are situations in which global constraints must be satisfied. For instance, in the link adaptation described in Sect. 6, the transmitter attempts to optimize its power efficiency while at the same time satisfying a global constraint of the link quality.

3. Individual radio has either partial view of the global state of the environment or limited resource to solve the problem. For instance, supposed that three radios are trapped in the underground and only one of them still has connectivity to the infrastructure which is located above ground. In order for the other two radios to find a route to connect back to the infrastructure, each of them must obtain the routing information of its neighbors in order to adjust its own routing policy.

In summary, collaboration plays an important role in cognitive radio adaptation. It helps an agent to obtain non-local information from other agents and have a complete view of the global environment. Based on such more complete information, an agent is able to better achieve the adaptation goal. Despite its benefit, collaboration incurs communication costs such as processing delay or increased overhead, which should not outweigh the performance improvement. In Sect. 10.3, we discuss a use case on which we evaluate the benefits and costs of collaboration adaptation.

Figure 3.3 shows how two radios exchange information about their knobs and meters. Assume that Radio A on the left hand side is the transmitter, and Radio B on the right hand side is the receiver. These two radios can exchange data and/or control messages, which include the knobs and meters. For example, the receiver can request a copy of the knobs and meters from the transmitter (shown in solid line and dash line with double arrows). It can also request the internal knobs and meters of its own (shown in solid line and dashed line with single arrows). Based on this

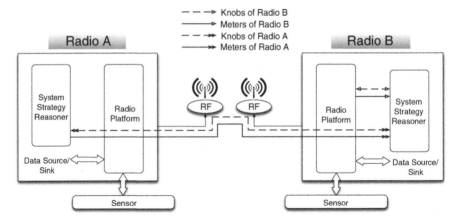

Fig. 3.3 Flows of knobs and meters between two cognitive radios

information, the SSR of the receiver will run an algorithm and select a configuration that will optimize the link performance. Then the SSR will send a request to its own radio platform, asking it to change the parameters accordingly. In addition, the receiver can also send a request to the SSR of the transmitter, requesting it to change its parameters accordingly.

Chapter 4
Signaling Options

To achieve collaborative adaptation discussed in Sect. 3.4, the adaption process will need to collect information from other radios and work with them to achieve the adaptation goal. This requires (1) a proper way to exchange control information (control messages, also referred to as *signaling*) between the radios, and (2) a proper way to interpret and execute the incoming control messages. The control messages must be capable of expressing more aspects than the current protocols can provide. For instance, instead of querying for a scalar parameter, cognitive radio should also be able to inquire for more complicated information, such as the structure of a radio component or the finite state machine of a component. Moreover, the way to interpret and execute the incoming control messages is expected to be flexible and efficient. In this chapter, we compare different signaling options and evaluate the flexibility of each option.

4.1 Fixed Protocol Vs. Flexible Signaling

There are three possible ways to achieve collaborative adaptation (Li et al. 2011).

1. *Fixed Protocol*: The first way would be to develop a communications protocol that is capable of expressing a wide range of aspects in wireless communications (flexible signaling). The extended flexibility of the protocol would result in the increase of the size of the header of the physical layer packets. However, it would still be limited by the size of the header and the types of information that could be included in the header. Additionally, since it is not possible to anticipate all the future needs at the design time, the coverage of the possible message types would still be limited.

2. *Flexible Signaling (XML-encoded messages)*: The second way would be to define a large vocabulary of control messages expressed in XML and include such messages in the payload of the packet. This approach provides more flexibility in that it can express more complicated signaling information, however, it would

S. Li and M.M. Kokar, *Flexible Adaptation in Cognitive Radios*,
Analog Circuits and Signal Processing, DOI 10.1007/978-1-4614-0968-7_4,
© Springer Science+Business Media, LLC 2013

require an XML schema to provide the description of the XML structure and procedural code to interpret the control messages written in XML.

3. *Flexible Signaling (OWL-encoded messages)*: The third approach would be to give radios a formal language with computer-interpretable semantics in which any control message can be encoded, provided that it can be expressed in terms of ontology shared by the radios. This approach does not require a separate procedural code to interpret each type of control messages; instead it requires a generic interpreter, i.e., an inference engine (reasoner) to process the control messages written in a formal language such OWL (Web Ontology Language) (Schreiber and Dean 2004) or RDF (Resource Description Framework) (RDF Working Group 2004).

Compared to the first approach, the XML and OWL approaches are both very flexible in terms of the number of possible message types. Practically, there is no limitation of what type of messages can be exchanged. When we need to make a change, in the XML approach one has to modify two things: the procedural code to process the XML file and the XML schema. In contrast, the OWL approach only needs the change of the ontology shared by the radios. In terms of inference capabilities, XML only has formal syntax but does not have formal semantics, therefore it cannot be processed by inference engines. Conversely, OWL has both formal syntax and formal semantics and therefore can be processed by the inference engines.

In conclusion, flexible signaling using OWL-encoded messages can provide a lot of flexibility to the existing protocols, i.e., an existing protocol can be extended by including an OWL-encoded control message in the payload of the packet without much change of the preamble frame structure. This type of signaling is used in the flexible collaborative adaptation discussed in this book.

4.2 Signaling Message Format and Overhead

The size of a control message depends on the message format. In general, there are two types of message format. The first one is the bit-oriented message format. One example of this type is Variable Message Format (VMF) (Rhyne et al. 2002; Priestnall 2010). In VMF, each protocol field is encoded as a binary number. The meaning of each binary number is given in the Data Element Dictionary (DED). The DED provides a mapping of a binary number to its represented information. In order to process an incoming VMF control message, the DED must be shared by the transmitter and receiver. Due to its compact format, this approach has small control message overhead, does not require much processing time and is most suitable for the bandwidth-constrained environments. However, the shortcoming of this approach is that VMF is not able to process the incoming messages with self-imposed errors, thus it has to retransmit the messages if errors occur. Additionally, any update of VMF requires a modification of the DED, thus increasing the cost of updating.

The second format is the character-oriented message format. Examples of this approach include Simple Object Access Protocol (SOAP) and the ontology and policy based approach used in this book. In this approach, a control message is encoded in XML or OWL/RDF (which is also serialized in XML). Due to the verbosity of XML, the sizes of the control messages are larger and thus it requires more processing time than some other approaches.

To get an intuition on the efficiency of this approach consider the following example. Suppose we need to send an integer number $4e9$. If it is encoded in binary format, it only requires $\frac{1}{8}log_2(4e9) = 4$ bytes. However, if it is encoded in text format, then each digit is viewed as a character. Assume each character is encoded using Extended ASCII (8 bits per character). Then it requires $8log_{10}(4e9) = 80$ bytes. The size of XML file can be reduced by using compression methods, e.g. Gzip or XMill.

4.3   Flexibility of Signaling Plans

As we discussed in Sects. 4.1 and 4.2, the implementation of control information, also referred to as *signaling*, has several options. From the perspective of the physical layer, control information can be either included in the protocol-defined preamble or in the extensible payload, shown in Fig. 4.1.

If the control information is included in the protocol-defined preamble, then the format of the control information is usually bit-oriented because the physical layer is the lowest layer and responsible for transmitting raw bits. The length, the ordering and the selection of the bit-oriented control information can be either fixed

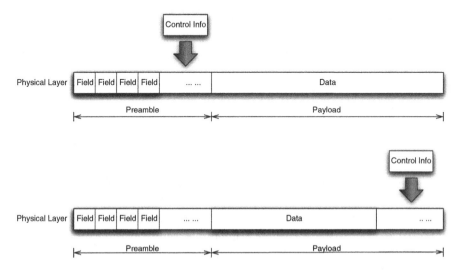

Fig. 4.1 Implementation of control information: in the preamble (*top*) vs. in the payload (*bottom*)

or variable. This kind of control messages have been used by the military Command & Control systems. For instance, HERIKKS[1] is an Early Warning Ground System in which the internal communications from a ground radar to a fire support unit uses control information of fixed length and fixed ordering (Sayin 2003). On the contrary, US VMF (Priestnall 2010; Rhyne et al. 2002) uses control information of variable length and ordering. The control information in the US VMF can be included in or omitted from the preamble as required (Sayin 2003). In both of the above cases, the length of the control information is bounded by the maximum length of the preamble.

If the control information is included in the extensible payload (in this case, the control information is also referred to as *control message*), then the format of the control information can be either bit-oriented or character-oriented because the control information can be inserted at the physical layer or any upper layer above (e.g. application layer). For example, SOAP (Simple Object Access Protocol) is an application protocol that uses XML to construct request and reply for the communications between the client and server (Tanenbaum 2002; Gudgin et al. 2007). Conversely, OBR uses OWL to construct the control information and the control information can be inserted at any upper layer. Both SOAP and OBR use character-oriented format for the control information.

We will compare all the above options with respect to the following aspects:

1. Protocol Extensibility, i.e. whether we can add additional types of control information to an existing protocol without changing the preamble frame structure.
2. Expressiveness, i.e. whether the control information is capable of expressing a wide range of types of control information.
3. Human-readability, i.e. whether the control information is human readable.
4. Ease of modification, i.e. how much needs to be modified when changes are needed.
5. Length of control information, i.e. whether the length of control information is fixed or variable.
6. Bounds of control information, i.e. whether there is an upper limit on the length of the control information.
7. Ordering of control information, i.e. whether the ordering of the control information is fixed or variable.
8. Suitability in bandwidth-constrained environment, i.e. whether the size of the control information small enough to be feasible in bandwidth-constraint environments.

Table 4.1 summarizes the comparison of the above options.

It can be seen that the ontology and policy based approach brings a great flexibility to the communications in the following aspects:

1. *Protocol extensibility*. We can extend any of the existing protocols with additional functionalities by including an OWL-encoded control message in the

---

[1] Hava Savunma Erken İkaz ve Komuta Kontrol Sistemi/Air Defense Early Warning System.

Table 4.1 Flexibility comparison

	Control information included in protocol-defined preamble		Control information included in extensible payload	
	Bit-oriented format (fixed)	Bit-oriented format (variable)	Character-oriented format (XML)	Character-oriented format (OWL)
Example	HERIKKS	US VMF	SOAP	OBR
Protocol extensibility	No	No	Yes	Yes
Expressiveness	Limited	Limited	Medium	Rich
Human readability	No	No	Yes	Yes
Ease of modification	– Protocol – Procedural code – Data Element Dictionary	– Protocol – Procedural code – Data Element Dictionary	– Procedural code – XML schema	– Ontology – Rules
Length of control information	Fixed	Variable	Variable	Variable
Bounds of the length of control information	Bounded to the maximum length of preamble	Bounded to the maximum length of preamble	Not bounded	Not bounded
Ordering of control information	Fixed	Variable	Variable	Variable
Suitability in bandwidth constrained environment	Suitable	Suitable	Not Suitable	Suitable

payload of the packet. There is no need to change the preamble frame structure because the added control message is included in the payload of the packet rather than hard-coded in the preamble.

2. *Expressiveness.* There is no limitation of what type of message can be exchanged because the control message is written in OWL. OWL is a highly expressive language which is capable of expressing a wild range of knowledge.

3. *Human-readability.* The OWL-encoded control message is human readable, i.e., even an XML representation is readable, but additionally, OWL can be viewed graphically in some of the OWL editing tools.

4. *Ease of modification.* We can easily change the behavior of the radio (e.g., how to respond to an incoming control message or how to react in a particular situation) by modifying the shared ontology (T-Box) and the policies (R-Box). There is no need to re-design the protocol such as the case of the fixed-protocol approach, or modify the procedural code and XML schema in the case of XML-encoded control message.

Table 4.2 Comparison of expressiveness of XML and OWL

		XML	OWL
Namespace		Yes	Yes
Class	Necessary and sufficient conditions	No	Yes
	Inheritance	No	Yes
	Instance	No	Yes
	Complement of class, union and intersection of classes	No	Yes
Property/ element constraints	Cardinality	Yes	Yes
	Domain/range	Yes	Yes
	Functional, inverse, symmetric, transitive	No	Yes
Datatypes		Yes (XML schema datatypes)	Yes (XML schema datatypes)

5. *Control Message.* Since OWL-encoded control messages are included in the extensible payload, the length of OWL-encoded control messages is not bounded to the length of the preamble. If the length of the control message exceeds the maximum length of the payload, it will be framed into a few smaller pieces and reassembled at the receiver. Also, the ordering of control information is flexible.
6. *Suitability in bandwidth constrained environment.* The use of OWL-encoded message will increase the overhead because OWL is verbose. However, the inference capability of OWL reduces the need for transmitting control information when it can be inferred locally. For this reason, we claim that OWL is suitable even for relatively bandwidth constrained environments.

To further evaluate expressiveness, we compare XML and OWL in the following aspects (Gil and Ratnakar 2002; Ferdinand et al. 2003; Walmsley and Fallside 2004; Thompson et al. 2004; Bray et al. 2008; Schreiber and Dean 2004; W3C OWL Working Group 2009):

1. Namespace, i.e. whether the language is capable of specifying namespaces.
2. Class, i.e., whether the language is capable of specifying: necessary and sufficient conditions for individuals to be classified as instances of a given class, inheritance, complement of a class (negation), unions and intersections of classes.
3. Property, i.e., whether the language is capable of specifying: domains and ranges of properties, cardinality (how many instances of the range can be associated via a property with an instance of the domain of the property), and some special types of property (functional, inverse, symmetric, or transitive).
4. Datatypes, i.e., whether the language supports datatypes.

The comparison of XML and OWL in terms of expressiveness is shown in Table 4.2. It can be seen that OWL is more expressive than XML and therefore capable of expressing a wide range of domain knowledge, such as classes and various types of property.

Chapter 5
Agent Communication Language

In the world of artificial intelligence, an agent needs to have a *representation model* to represent itself and its view of the world. A representation model includes ontologies, which are vocabularies and taxonomy that define the basic terms and relationships in a given domain, and a content language to represent these terms and relationships. For instance, in the domain of wireless communications, agents need to have vocabularies to represent various components (e.g. equalizer, antenna, etc.); processing (e.g. sampling, coding/decoding, frequency spreading, etc.) and goals of actions (e.g. maximizing throughput or minimizing interference). Given such a representation model, agents also need a *communication model* to capture the communications and flow of knowledge exchange within the agent community. A communication model requires a set of language primitives that can be used to implement the model. An Agent Communication Language (ACL) is commonly known as a language that provides a set of language primitives to implement the agent communication model. ACLs are only used to construct the *message wrapper* and not concern about the choice of the content language and the ontology model (Vasudevan 1998).

Since agents might act on behalf of human, agent communication are in many ways similar to human communication. In human society, people mainly communicate with each other using speech, thus it is natural to derive the ACL language primitives from the linguistic theory of *speech acts*. One of the most influential work on speech acts is John Austin's book *How to Do Things with Words* published in 1962 (Austin 1962). In this book, he argues that a sentence may not describe something or state true or false, instead, the uttering of such sentence is part of doing an action, which he calls *performative utterance* (Austin 1962, p. 6). The action which is performed when a performative utterance is issued belongs to what he calls *speech act*. In his doctrine, speech acts can be classified into: locutionary act, illocutionary act and perlocutionary act. *Locutionary act* is an act *of* saying something (Austin 1962, p. 94); *illocutionary act* is an act *in* saying something (Austin 1962, p. 99); *perlocutionary act* is an action performed by saying something (Austin 1962, p. 101). For instance, David said to me "Shoot her", the uttering of this sentence can be a locutionary act of David saying "shoot her"; or can be an

S. Li and M.M. Kokar, *Flexible Adaptation in Cognitive Radios*,
Analog Circuits and Signal Processing, DOI 10.1007/978-1-4614-0968-7_5,
© Springer Science+Business Media, LLC 2013

Table 5.1 Classification of illocutionary speech acts

Type	Description	Examples
Representatives	Commit the speaker to the truth of the expressed proposition	Conclude, deduce, etc.
Directives	Attempts by the speaker to get the hearer to do something	Order, command, request, etc.
Commissives	Commit the speaker to some future course of action	Promise, vow, pledge, guarantee, swear, etc.
Expressives	Express the psychological state in the propositional content	Thank, congratulate, apologize, condole, welcome, etc.
Declarations	Bring alternation in the status or condition of the referred-to objects if the declaration has been successfully performed	"I declare your employment is terminated", "I pronounce you husband and wife"

illocutionary act that David urged (or advised, ordered, etc.) me to shoot her; or can be a perlocutionary act that David persuaded me to shoot her. Another important work on speech act is John Searle's classification of illocutionary acts which later becomes the foundation of ACL. He proposed five basic types of illocutionary acts, which are representatives, directives, commissives, expressives, and declarations, shown in Table 5.1

The theory of speech acts classifies the human (or machine) utterance into several categories based on the intention of the speaker, the impact on the hearer and any other physical manifestations of the utterance (Vasudevan 1998). This classification provides the foundation to design and develop the language primitives of ACL. The most popular ACLs are KQML (Knowledge Query and Manipulation Language) (Finin et al. 1993) and FIPA (Foundation for Intelligent Physical Agents) ACL (FIPA 2002c). They both serve as the wrapper language of providing the language primitives to implement the agent communication models and is unaware of the choice of the content language. However, in KQML, an agent can have primitives such as "*insert*" to directly manipulate another agent; whereas FIPA-ACL does not have such primitives. Instead FIPA-ACL has primitives such as "*request*" to indirectly manipulate another agent. In other words, there is a fundamental difference between KQML and FIPA-ACL in terms of "what is done in the wrapper language and what is delegated by the wrapper language to the content language" (Vasudevan 1998).

In this book, we are going to solely focus on FIPA-ACL. The FIPA-ACL is a set of specifications that covers: (1) an ACL message structure that can be used by agents to construct messages; (2) a set of communicative acts (language primitives) that specifies the types of ACL messages; (3) a set of communication protocols that support the message exchange interaction. In other words, FIPA-ACL is a set of specifications to support interoperation of heterogeneous agents and services. It defines the envelop of the message and the protocols to support message exchange. Specific implementation is free to choose or define its own language to represent the content of the message, e.g. XML, OWL, etc. In addition, FIPA-ACL does not provide implementation details of how to implement the interaction protocols and how to interpret and execute an ACL message. These decisions are left to specific implementation.

In the rest of this chapter, we will first introduce the FIPA-ACL message structure, the library communicative acts and show concrete examples of the FIPA-ACL protocols.

5.1 FIPA-ACL Message Structure

A standardized message structure provides a foundation for agents to construct message and exchange message.

A FIPA-ACL message consists of two parts: envelop and content. The envelop of the message is constructed using a set of message parameters. The message parameters provide information such as the type of the message, the sender and receiver, the conversation id, etc. The content of the message is expressed using a content language. Available options of FIPA content language include Semantic Language (SL) (FIPA 2002k), Constraint Choice Language (CCL) (Willmott et al. 1999), Knowledge Interchange Format (KIF) (Genesereth and Fikes 1992), Resource Description Framework (RDF) (RDF Working Group 2004), Web Ontology Language (OWL) (Schreiber and Dean 2004; W3C OWL Working Group 2009), etc. The content language is expected to have formal syntax and semantics. The sending agent and receiver should provide a mechanism to interpret the content of the message.

Table 5.2 shows a complete list of FIPA-ACL message parameters. Not all the parameters in this table are mandatory in FIPA-ACL message. *performative* is the

Table 5.2 FIPA-ACL message parameters (*source*: FIPA 2002a)

Parameter	Description
Performative	ACL message type
Sender	Sender of the message
Receiver	Receiver of the message
Reply-to	Receiver of the subsequent messages in this conversation thread
Content	Content of message
Language	Language in which the content is expressed
Encoding	Method in which the content is encoded
Ontology	Ontology(s) used to support interpretation of the content
Protocol	Interaction protocol in which the ACL message is generated
Conversation-id	Identifier used to identify the ongoing sequence of communicative acts that together form a conversation
Reply-with	An expression that will be used by the responding agent to identify This message
In-reply-to	An expression that refers an earlier action to which this message A reply
Reply-by	Time or date indicate the latest time by which the sending agent would Like to receive a reply

```
(request
      :sender Tom
      :receiver Anna
      :content (
            <?xml version="1.0"?>
            <rdf:RDF xmlns:rdf="http://www.w3.org/1999/02/22-rdf-syntax-ns#"
                  xmlns:fipa="http://www.fipa.org/schema#">
              <fipa:Action rdf:ID="TomAction1">
                <fipa:actor>Anna</fipa:actor>
                <fipa:act>close</fipa:act>
                <fipa:argument>window</fipa:argument>
              </fipa:Action>
            </rdf:RDF>
      )
      :language fipa-rdf
      :reply-with request1
)
```

Fig. 5.1 Example of FIPA-ACL message: request

```
(inform
      :sender Anna
      :receiver John
      :content (
            <?xml version="1.0"?>
            <rdf:RDF xmlns:rdf="http://www.w3.org/1999/02/22-rdf-syntax-ns#"
                  xmlns:fipa="http://www.fipa.org/schema#">
              <fipa:Action rdf:about ="TomAction1">
                <fipa:done>true</fipa:done >
                <fipa:result>window closed</fipa:result >
              </fipa:Action>
            </rdf:RDF>
      )
      :language fipa-rdf
      :in-reply-to request1
)
```

Fig. 5.2 Example of FIPA-ACL message: inform

only parameter that is required in the FIPA-ACL message. Which parameters will be chosen to construct the message depends on specific application. For instance, Fig. 5.1 is an example of FIPA-ACL message. The type of this message is *request*, which is shown as *performative* at the beginning of the message. The sending agent is *Tom* and the receiving agent is *Anna*. The content is expressed in RDF, which is the shown in the *language* field. The content says: "Tom requests Anna to close the window". In reply to this request, Anna sends an *inform* message to Tom confirming that the window is closed. The *inform* message is shown in Fig. 5.2

5.2 Communicative Act Library

FIPA-ACL specification defines a Communicative Act Library that consists of 22 types of communicative acts. This library is developed based on the speech act theory, shown in Table 5.3 (FIPA 2002c).

Table 5.3 Types of FIPA communicative act (*source*: FIPA 2002c)

	Communicative act	Description
1	Accept proposal	Agent i accepts a previously proposal submitted by agent j to perform an action
2	Agree	Agent i agrees to perform some actions
3	Cancel	Agent i informs agent j that agent i no longer has the intention that agent j performs some actions
4	Call for proposal	Agent i calls for proposals to perform a given action ("Call for proposal" is usually used to initiate a negation process)
5	Confirm	Agent i informs agent j that a given proposition is true, where agent j is uncertain about the proposition
6	Disconfirm	Agent i informs agent j that a given proposition is false, where agent j believes the proposition is true
7	Failure	Agent i tells agent j that agent i attempted to perform an action but failed
8	Inform	Agent i informs agent j that a given proposition is true
9	Inform If	Agent i informs agent j that whether or not a proposition is true ("Inform if" is usually used with "Query If" as a pair)
10	Inform Ref	Agent i informs agent j some object that corresponds to a descriptor, such as a name ("Inform Ref" is usually used with "Query Ref" as a pair)
11	Not understood	Agent i tells agent j that agent i did not understand what agent j just did (a typical example is that agent i did not understand the message agent j just sent to agent i)
12	Propagate	Agent i sends a message to agent j and wants agent j to identify the agents denoted by the given descriptor and send the message to those agents
13	Propose	Agent i submits a proposal to perform an action to agent j
14	Proxy	Agent i requests agent j to identify agents that satisfy the given descriptor and sends the embedded message to those agents
15	Query If	Agent i asks agent j whether or not a given proposition is true
16	Query Ref	Agent i asks agent j for the object identified by a descriptor
17	Refuse	Agent i refuses agent j to perform a given action and explains the reason for the refusal
18	Reject proposal	Agent i rejects a proposal to perform some action during a negotiation
19	Request	Agent i requests agent j to perform some action
20	Request when	Agent i wants agent j to perform some action when some given proposition becomes true
21	Request whenever	Agent i wants agent j to perform some action as soon as some given proposition becomes true and thereafter each time the proposition becomes true again
22	Subscribe	Agent i requests a persistent intention to be notified of the value of a reference and to be notified again whenever the object identified by the reference changes (the "subscribe" act is terminated by a "cancel" act)

5.3   Interaction Protocols

FIPA-ACL defines nine interaction protocols, which are summarized in Table 5.4

Each protocol is described using a UML sequence diagram, shown in Fig. 5.3 shows an example of the Request Interaction Protocol. At any point in the Interaction Protocol, the receiver agent can inform the sender agent that it does not understand the message that the sender just sent to the receiver. In this case, the receiver will send a *not-understood* to the sender. In addition, in any point in the Interaction Protocol, the initiator agent can cancel the interaction by sending a *cancel* message to the participant agent, following the meta protocol shown in Fig. 5.4 (FIPA 2002i).

Table 5.4 FIPA interaction protocols (*source*: FIPA 2002b,d,e,f,g,h,i,j,l)

Interaction protocol (IP)	Description
Query IP	Agent *i* asks agent *j* whether or not a given proposition is true, or asks agent *j* for the object identified by a descriptor
Request IP	Agent *i* requests agent *j* to perform some actions
Request When IP	Agent *i* wants agent *j* to perform some action when some given proposition becomes true
Contract Net IP	Agent *i* (the Initiator) calls for proposals from other agents (the Participants) to perform a given task
Iterated Contract Net IP	Iterated Contract Net IP is an extension of Contract Net IP, it allows multi-round iterative negotiation
Brokering IP	Agent *i* requests a broker agent *j* to find one or more agents who can answer a query. The broker *j* then determines a set of appropriate agents to which to forward the query, sends the query to those agents and relays their answers back to the original requestor *i*. (A broker is an agent that offers a set of communication facilitation services to other agents using some knowledge about the requirements and capabilities of those agents)
Recruiting IP	Recruiting IP is similar to Brokering IP, the only difference is that the answers from the selected target agents either go directly back to the original requestor *i* or to some designated receivers.
Subscribe IP	Agent *i* requests agent *j* to perform an action on subscription and subsequently when the referenced object changes
Propose IP	Agent *i* (the Initiator) proposes to agent *j* that *i* will do the actions described in the *propose* communicative act when *j* accepts the proposal

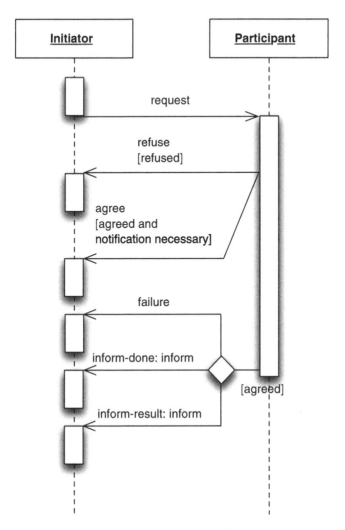

Fig. 5.3 FIPA request interaction protocol (*source*: FIPA 2002i)

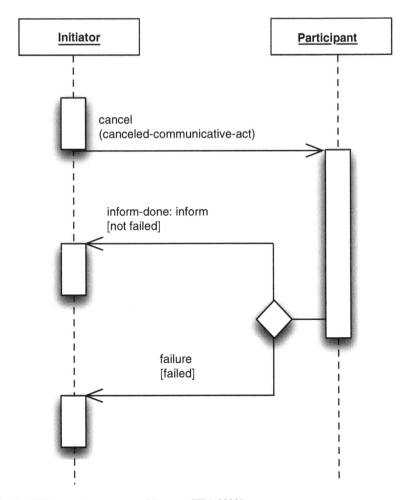

Fig. 5.4 FIPA cancel meta protocol (*source*: FIPA 2002i)

Chapter 6
An Example: Collaborative Link Adaptation

A wireless communications link consists of a transmitter–receiver pair, and the wireless medium via which information is transferred. The general goal of link adaptation is to *maximize the information bit rate per transmitted watt of power* subject to a set of constraints. This is attained by fine-tuning the parameters in the transmitter and the receiver, while the channel parameters are assigned with approximate values by estimation.

6.1 Description of Communications Parameters

First, we will take a look at the parameters of the transmitter and receiver.

6.1.1 Transmitter Parameters

- *payloadsize*

This is the size of message field plus control field. This is referred to as the payload, and is passed to the CRC encoding routine.

- *m*

This is the integer index for the $(2^m - 1, 2^m - 1 - m)$ Hamming code. That is, the number of information bits per codeword is $2^m - 1 - m$, and the codeword length is $2^m - 1$. The index must be 2 or higher. Note that the coding overhead is given by: $m/(2^m - 1)$. Since a Hamming decoder can only correct a single error in $2^m - 1$ received bits, as m increases, the ratio between the size of overhead and the size of the whole packet decreases. There is no natural upper bound of m.

S. Li and M.M. Kokar, *Flexible Adaptation in Cognitive Radios*,
Analog Circuits and Signal Processing, DOI 10.1007/978-1-4614-0968-7_6,
© Springer Science+Business Media, LLC 2013

- *trainPeriod*

This is the length of the training sequence, in channel symbols.

- *fracSpacing*

This is the number of samples per channel symbol. It is usually not changed.

- *v*

This is the positive integer which controls the size of the QAM constellation, which is 4^v. The number of coded bits per symbol is $2v$.

- *PowdB*

This is the transmission power measured in dBm.

6.1.2   Receiver Parameters

- *M*

This the positive integer number of feedback taps in the equalizer.

- *N1*

This is the positive integer number of precursor feedforward taps.

- *N2*

This is the positive number of postcursor feedforward taps. The general rule for specifying M, $N1$, and $N2$ are:

$$N1 + N2 = length(mdp) \tag{6.1}$$

$$M = \frac{N1 + N2}{fracSpacing} \tag{6.2}$$

Here, *mdp* is the multipath vector. Larger values of M, $N1$ and $N2$ are also acceptable, but the length of the shortest training sequence is approximately $5 * (N1 + N2 + M)$.

- *fracSpacing*

This is the number of (receiver) samples per symbol. It is usually set to 2.

- *Memory*

This is the number of samples over which the channel is assumed to be constant. It is used to train the equalizer coefficients. It should never exceed $\frac{cohtime}{5}$ (*cohtime* is the coherence time over which a propagation wave may be considered constant), as the channel has changed in this window. A smaller value yields a more nimble equalizer, but yields a smaller equalizer SNR. The goal would be to set:

$$Memory \geq \frac{cohtime}{10} \tag{6.3}$$

and keep the SNR as high as possible.

- *mSNR*

This is the reported equalizer SNR, in dB. Intuitively, a value greater than 10 dB yields good detection performance, but a value greater than 15 dB indicates that the data rate could be increased, or the transmit power should be decreased, etc.

6.1.3 Parameters Summary

The summary of all the above parameters is shown in Table 6.1.

Table 6.1 Parameters summary

No.	Transmitter parameters	Receiver parameters	Unit	Notes	Constant/ meter/knob
1	payloadsize		Byte	Payload = message field + control field	Constant (=128)
2	m		N/A	(2^m-1, 2^m-1-m) hamming code	Knob
3	trainPeriod		Number of channel symbol		Knob
4	fracSpacing	fracSpacing	Number of samples per channel symbol		Constant (=2)
5	v		N/A	4^v is the size of QAM constellation	Knob
6	PowdB		dB		Knob
7		M	N/A	Number of feedback taps in equalizer	Knob
8		N1	N/A	Number of precursor feedforward taps	Knob
9		N2	N/A	Number of postcursor feedforward taps	Knob
10		Memory	Number of sample		Knob
11		mSNR	dB		Meter

+

6.2 Objective Function

In this link adaptation problem, the goal is to maximize the information bit rate per transmitted watt of power. The computation of information bit rate is shown as follows.

The payload size is fixed per packet to be $payloadsize \cdot 8$ information bits. "Information bits" means the bits comprising the message, the control field, and any padding. These bits are passed into a CRC32 checker which postpends 32 bits. The result is then coded in the following way:

- $(32 + payloadsize \cdot 8)$ is padded so that the number of bits is evenly divided by $2^m - m - 1$. We neglect this in the calculation.
- The bit stream is coded to yield approximately

$$(32 + payloadsize \cdot 8)(1 + \frac{m}{2^m - m - 1}) \tag{6.4}$$

coded bits.

- The coded bit stream is QAM modulated to form

$$\frac{(32 + payloadsize \cdot 8)(1 + \frac{m}{2^m - m - 1})}{2 \cdot v} \tag{6.5}$$

- QAM channel symbols. Again, a few additional bits are postpended to make the new length divisible by $2v$.
- QAM symbols are prepended to form the training sequence. The total number of QAM symbols in the packet is

$$trainPeriod + \frac{(32 + payloadsize \cdot 8)(1 + \frac{m}{2^m - m - 1})}{2 \cdot v} \tag{6.6}$$

- The transmitter uses

$$\frac{10^{\frac{PowdB}{10}}}{1000} \cdot \frac{[trainPeriod + \frac{(32 + payloadsize \cdot 8)(1 + \frac{m}{2^m - m - 1})}{2 \cdot v}] \cdot fracSpacing}{sampleRate} \tag{6.7}$$

Joules of energy to send $payloadsize \cdot 8$ bits.

- The goal is to maximize the information bit rate per transmitted watt of power, hence the metric to maximize is

$$f = \frac{payloadsize \cdot 8 \cdot sampleRate}{\frac{10^{\frac{PowdB}{10}}}{1000} \cdot [trainPeriod + \frac{(32+payloadsize \cdot 8)(1+\frac{m}{2^m-m-1})}{2 \cdot v}] \cdot fracSpacing}$$

(6.8)

Suppose $sampleRate$, $payloadsize$ and $fracSpacing$ are fixed, then there are four variables in the objective function: $PowdB$, $trainPeriod$, m and v. The increase of $PowdB$ or $trainPeriod$ will produce an increase of the objective function. The increase of v or m will yield to a decrease of the objective function. Also, $PowdB$, $trainPeriod$, and v affect the value of another variable $mSNR$, which will be discussed in the following section. The range of $mSNR$ must be between 10 to 15.

6.3 Constraints

Suppose for the nth transmission, $PowdB_n$, $mSNR_n$ and v_n are the transmission power, signal-to-noise ratio, and the size of the QAM constellation, respectively.

1. The reported equalizer SNR_n must be between 10 and 15 dB. Intuitively, a value greater than 10 dB yields good detection performance, but a value greater than 15 dB indicates that the data rate could be increased, or the transmit power should be decreased. Hence, the constraints for $mSNR_n$ is:

$$10 \leq mSNR_n \leq 15$$

(6.9)

2. $PowdB$ is the transmit power in dB. Here, we set the upper bound of $PowdB$ as:

$$PowdB \leq 0 dB$$

(6.10)

3. Suppose

$$\Delta PowdB_n = PowdB_n - PowdB_{n-1}$$

(6.11)

and

$$\Delta mSNR_n = mSNR_n - mSNR_{n-1}$$

(6.12)

Since both $PowdB$ and $mSNR$ are in dB, a drop of $PowdB$ results in an equal drop in $mSNR$. Thus

$$\Delta PowdB_n = \Delta mSNR_n$$

(6.13)

To guarantee Eq. (6.9), $\Delta mSNR_n$ must not exceed $15 - mSNR_{n-1}$ and not be less than $10 - mSNR_{n-1}$. Hence,

$$10 - mSNR_{n-1} \leq \Delta PowdB_n \leq 15 - mSNR_{n-1} \tag{6.14}$$

that is:

$$10 - mSNR_{n-1} + PowdB_{n-1} \leq PowdB_n \leq 15 - mSNR_{n-1} + PowdB_{n-1} \tag{6.15}$$

4. The parameter m is the integer index for the $(2^m - 1, 2^m - 1 - m)$ Hamming code. That is, the number of information bits per codeword is $2^m - 1 - m$, and the codeword length is $2^m - 1$. The index must be 2 or higher. Thus the lower bound of m is 2. The parameter m does not effect the equalizer's SNR, as it controls the coding overhead. There is no natural upper bound of m. However, since the length of the overhead must be larger than zero, we can compute an approximate upper bound of m by assuming length of the payload is fixed. In the MATLAB simulation, $payloadsize$ is fixed to 128 bits, according to the discussion in Sect. 6.2, the length of the Hamming code overhead equals to:

$$(payloadsize \cdot 8 + 32) \cdot \frac{m}{2^m - 1 - m} = 1056 \cdot \frac{m}{2^m - 1 - m} \tag{6.16}$$

Here we set

$$m \leq 10 \tag{6.17}$$

Hence the lowerbound of the length of the Hamming code overhead approximately equals to 10.

5. The parameter v controls the size of the QAM constellation, the natural lower bound of v is:

$$v \geq 1 \tag{6.18}$$

Parameter v does affect equalizer performance, in the following way. For a given value of v, the QAM constellation has a maximum magnitude of unity, achieved at the corners. There are 4^v points uniformly in a rectangular grid, and the minimum distance between distinct constellation points is $\frac{1}{\sqrt{2}(2^v-1)}$. Consequently, a possible increase in v by 1 unit would drop the SNR by the factor $(\frac{2^v-1}{2^{v+1}} - 1)^2$, or approximately by $\frac{1}{4^v}$, which is 6 dB. [1] In short, increasing v by one unit drops the equalizer SNR by approximately 6 dB. Suppose

$$\Delta v_n = v_n - v_{n-1} \tag{6.19}$$

---

[1] In fact, since $10log4v = 6 + 10logv$, the drop is $6 + 10logvdb$.

then

$$\Delta v_n = -\frac{\Delta m SNR_n}{6} \tag{6.20}$$

Again, to guarantee Eq. (6.9), $\Delta m SNR_n$ must not exceed $15 - mSNR_{n-1}$ and not be less than $10 - mSNR_{n-1}$. Hence,

$$\frac{mSNR_{n-1} - 15}{6} \le \Delta v_n \le \frac{mSNR_{n-1} - 10}{6} \tag{6.21}$$

This constraint can be further simplified to

$$\left\lceil \frac{mSNR_{n-1} - 15}{6} \right\rceil + v_{n-1} \le v_n \le \left\lfloor \frac{mSNR_{n-1} - 10}{6} \right\rfloor + v_{n-1} \tag{6.22}$$

6. The parameter $trainPeriod$ affects the equalizer performance in a less clear way. If $trainPeriod$ is less than $5 * (M + N1 + N2)$, then the equalizer does not fully converge. The QAM symbol detection may fail completely, or recover after an initial burst of symbol errors. Recall that our coding cannot handle error bursts, so if $trainPeriod$ is reduced below that critical value, CRC errors may suddenly appear. On the other hand, making $trainPeriod$ greater than twice the critical value will have little effect on equalizer performance, but will work against the maximization of the metric. Hence, the constraint of parameter $trainPeriod$ is:

$$5 \cdot (M_n + N1_n + N2_n) \le trainPeriod_n \le 10 \cdot (M_n + N1_n + N2_n) \tag{6.23}$$

7. Clearly, M, $N1$, $N2$ have a threshold influence on equalizer performance: the equalizer SNR will increase with M, $N1$, or $N2$, until a sufficiently large equalizer for the multipath is achieved. After that point, increasing the equalizer dimensions will have no effect, except to increase the shortest possible training sequence.

8. The equalizer SNR will increase with the parameter $Memory$. Then, it flattens out, and decrease as $Memory$ exceeds $\frac{cohtime}{5}$ as mentioned earlier. On the other hand, a smaller value yields a more nimble equalizer. Here, the range of $Memory$ is set to:

$$\frac{cohtime}{10} \le Memory \le \frac{cohtime}{5} \tag{6.24}$$

6.4 Summary of Objective Function and Constraints

The objective function and the constraints are stated in the following.

Objective Function and Constraints. Suppose $payloadsize$, $sampleRate$, and $fracSpacing$ are constants, $\{PowdB_{n-1}, trainPeriod_{n-1}, m_{n-1}, v_{n-1},$

Table 6.2 Constraints of the link adaptation problem

	Constraints
1	$10 \leq mSNR_n \leq 15$
2	$PowdB_n \leq 0$
3	$10 - mSNR_{n-1} + PowdB_{n-1} \leq PowdB_n \leq 15 - mSNR_{n-1} + PowdB_{n-1}$
4	$N1_n + N2_n \geq 4$
5	$M_n \geq \frac{N1_n + N2_n}{2}$
6	$5 * (M_n + N1_n + N2_n) \leq trainPeriod_n \leq 10 * (M_n + N1_n + N2_n)$
7	$\frac{cohtime_{n-1}}{10} \leq Memory_n \leq \frac{cohtime_{n-1}}{5}$
8	$2 \leq m_n \leq 10$
9	$1 \leq v_n$
10	$\lceil \frac{mSNR_{n-1}-15}{6} \rceil + v_{n-1} \leq v_n \leq \lfloor \frac{mSNR_n-10}{6} \rfloor + v_{n-1}$
11	$M_n, N1_n, N2_n, trainPeriod_n, Memory_n, m_n, v_n$ are integer

Notes:
1. $mSNR_n$ will increase with M_n, $N1_n$, or $N2_n$, until a sufficiently large equalizer for the multipath is achieved. After that point, increasing the equalizer dimensions will have no effect, except to increase the shortest possible training sequence
2. $mSNR_n$ will increase with the parameter $Memory_n$, flatten out, and then decrease as $Memory_n$ exceeds $\frac{cohtime}{5}$
3. $cohtime_{n-1}$ is a known parameter that is estimated at the receiver

$mSNR_{n-1}$, M_{n-1}, $N1_{n-1}$, $N2_{n-1}$, $Memory_{n-1}$ } are known knobs and meters obtained from the $n - 1$ th transmission. {$PowdB_n$, $trainPeriod_n$, m_n, v_n, $mSNR_n$, M_n, $N1_n$, $N2_n$, $Memory_n$ } are tunable knobs that will be adjusted for the n th transmission.

- Objective function

Maximize f [Eq. (6.8)]

- Subject to the constraints shown in Table 6.2

6.5   Query Vs. Estimation

The objective of the link adaptation problem is to maximize f [Eq. (6.8)]. The four parameters in this objective function are the tunable knobs of the transmitter. The constraints require that the measured $mSNR$ at the receiver must be between 10 and 15 dB. The $mSNR$ is affected by the transmitter knobs {$PowdB$, $trainPeriod$, v, m} and the receiver knobs {M, $N1$, $N2$, $Memory$}. From the perspective of transmitter, it attempts to maximize the objective function by adjusting its local parameters. However, due to the complicated dependencies among the transmitter knobs, the receiver knobs, and the meters measured at the receiver, the transmitter needs to gather the information at the receiver before it is able to make a decision of how to adjust its local parameters. From the perspective of the receiver, it plays a principle role of measuring the quality of the link. In addition, its local parameter also affects the data reception and link quality though they have no impact on the objective function. In other words, the problem we are trying to solve is to optimize the power efficiency at the transmitter end, yet both of the transmitter and receiver must collaborate to satisfy the global constraint of the link quality. Inherently, both the transmitter and receiver have only partial information of the global state, there are two options for them to obtain a complete view of the global state:

1. *Query for the desired information from others.* When a receiver gets a data packet, it will measure the $mSNR$. If the $mSNR$ does not satisfy the constraint, the receiver can query the transmitter for its local parameters. Similarly, a transmitter can also query for the $mSNR$ from the receiver. In other words, both of the transmitter and receiver can obtain a complete view of the global state by querying the desired information from others. Who will be in charge of the decision making is not our concern, i.e. we can either assign one radio as the master and let it make a decision on how to adjust the parameters on both ends, or we can let the transmitter and receiver negotiate and make a decision together. The benefit of this approach is that an agent has sufficient and accurate information to solve the problem by obtaining all the desired information from other agents. The shortcoming is that this approach incurs communications cost such as processing delay or increased overhead. In addition, the agents must establish a way to exchange, interpret and execute the information. This approach will benefit most in the situations when it is difficult to obtain a model to estimate non-local information.

2. *Estimate the desired information locally.* When a receiver gets a data packet, it will measure the $mSNR$. If the $mSNR$ does not satisfy the constraint, the receiver can use a model (if it is available) to estimate the parameters at the transmitter end, e.g. it can estimate the transmit power by the measured $mSNR$ and the assumptions made on the channel environment (e.g. path loss, channel type, noise level, etc.). In this approach, there is no communications cost because all the estimation and decision making are done locally. However, this approach only applies to the cases when the estimation model is available.

In the link adaptation problem, $mSNR$ has dependency on $\{PowdB, trainPeriod, v, m\}$ and $\{M, N1, N2, Memory\}$. It is difficult to express $mSNR$ in a close-form formula of all the above parameters, i.e. $mSNR = f(PowdB, trainPeriod, v, m, M, N1, N2, Memory)$. Therefore, it is more preferable to let the two radios exchange their local informations rather than doing estimation locally.

6.6 The Process of Link Adaptation

In this section, we will summarize the process of link adaptation. Assume that receiver is in charge of the decision making, then from the receiver's point of view, the percepts from the environment include the knobs of the transmitter and receiver as well as the meters obtained from the receiver. Table 6.3 summarizes the description of the percepts, actions, goals and environment of the receiver.

Table 6.3 The description of the percepts, actions, goals and environment for the radio agent in the link adaptation problem

Agent type	Receiver
Percepts	Knobs from the $n-1$th transmission: $\{PowdB_{n-1}, trainPeriod_{n-1}, m_{n-1}, v_{n-1}, M_{n-1}, N1_{n-1}, N2_{n-1}, Memory_{n-1}\}$ Meter from the $n-1$th transmission: $\{mSNR_{n-1}\}$
Actions	Decision on the new configurations of knobs for the nth transmission: $\{PowdB_n, trainPeriod_n, m_n, v_n, M_n, N1_n, N2_n, Memory_n\}$
Goals	Maximize the information bit rate per watt power [Eq. (6.8)]
Environment	Wireless channel

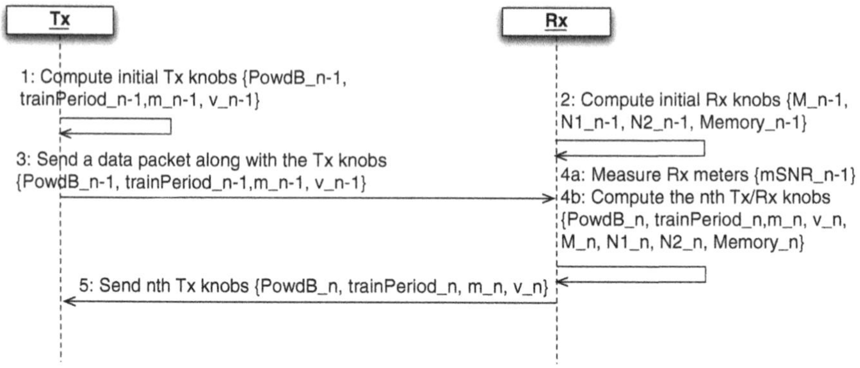

Fig. 6.1 Sequence diagram of collaborative link adaptation

The basic adaptation process for this link adaptation problem has the following steps:

1. In the $n-1$ th transmission, the values of the tunable transmitter parameters are $\{PowdB_{n-1}, trainPeriod_{n-1}, m_{n-1}, v_{n-1}\}$, and the values of the tunable receiver parameters are $\{mSNR_{n-1}, M_{n-1}, N1_{n-1}, N2_{n-1}, Memory_{n-1}\}$. Using this set of parameters, the transmitter sends a data packet to the receiver.
2. The receiver receives the data packet, and then runs an adaptation algorithm to compute the optimized values of the transmitter parameters and the receiver parameters for the n th transmission, i.e. $\{PowdB_n, trainPeriod_n, m_n, v_n, M_n, N1_n, N2_n, Memory_n\}$.
3. Then the receiver sends the suggested parameters values $\{PowdB_n, trainPeriod_n, m_n, v_n\}$ to the transmitter.
4. If the transmitter accepts these suggested values, it will change its transmission parameters accordingly. Otherwise, the transmitter will negotiate with the receiver and repeat steps 1–3 until they both agree on a new set of parameters values.

Figure 6.1 shows the sequence diagram of an example collaboration strategy between two radios.

Chapter 7
Knowledge and Inference

Ontology Based Radio (OBR) uses ontologies to represent its knowledge about itself and the communication environment. This knowledge is manipulated by generic components called *reasoners* or *inference engines*. The conceptual architecture of OBR in Chap. 2 includes two reasoners—System Strategy Reasoner (SSR) and Policy Conformance Reasoner (PCR). Since the term "reasoning" may mean different things to different people, in this chapter we elaborate on the way this term is used in this book.

In short, reasoning means here deriving facts from other facts using some *inference rules*. For instance, since it is common knowledge that every radio has an antenna, from the fact that a specific piece of equipment is a radio, one can infer that it has an antenna. While this is obvious to humans, for computers to be able to carry out this kind of inferences, they must be equipped with rules of inference and software that invokes this kind of processing. In this simple example, the inference rule could be stated informally as follows: "if X is an instance of class C and if every instance of class C has a feature F, then X has feature F." In this case C corresponds to the class of Radio, while feature F corresponds to the class Antenna. "has feature" is the name of a relation between instances of Radio and instances of Antenna. "is an instance" is another relation that relates instances of any class with the class.

More formally, such reasoning can be represented as the application of the rule like this:

$$(\forall c \in C \bullet p(c)) \wedge c_1 \in C \rightarrow p(c_1)$$

A reasoner has a number of such rules and applies them to the various information items (in this case to c_1) and derives new facts (in this case $p(c_1)$). This kind of reasoning is referred to as *logical reasoning* or *formal reasoning*.

The term "reasoning" is often used to mean something different, e.g., as a function written in a procedural language (e.g., Java) that takes input values of some sort (e.g., strings) and returns as values some other type of individuals, e.g., numbers. In this scheme the reasoning is said to be *procedural*. Formal reasoning, on the other hand, is *declarative*, in the sense that the knowledge is represented separately from the algorithms that process it. In other words, the issue then is

S. Li and M.M. Kokar, *Flexible Adaptation in Cognitive Radios*,
Analog Circuits and Signal Processing, DOI 10.1007/978-1-4614-0968-7_7,
© Springer Science+Business Media, LLC 2013

what should be the strategy—put more emphasis on declarative representation of knowledge as a separate entity or put knowledge into procedural code? The former approach is referred to as "knowledge-rich", while the latter as "knowledge-less". We analyze this issue in the following section.

7.1 Knowledge-Less Vs. Knowledge-Rich

To clarify the distinction between the knowledge-rich and knowledge-less approaches, it is useful to focus on the differences between data and knowledge. Without going into formal definitions, we can simply consider data, for instance, as information stored in databases. Thus a piece of data is a row in a table along with the associations of the data items with the columns of the table. So for instance if the table represents frequencies for particular transmissions, then a row would contain two data items $< t_i, 5 >$, where t_i would represent a particular transmission event and the number 5 would represent the frequency of 5 MHz. Another example would be the raw data of the environment collected through sensors, e.g. interference, battery life, position.

Data can be used as the input to an adaptation routine, which would then computed the values of specific radio knobs based on such input. We are striving for solutions that are generic, rather than applicable only to a very narrow types of input data and processing requirements. One way to achieve such a generic capability is by using search as a technique for problem solving, rather than multiple problem specific algorithms. This approach is termed sometimes as "generate-and-search". In this approach, a generic search algorithm is used to search through a *space* of potential solutions; the visited solution points in such a space are then checked whether they satisfy the requirements of the solution or whether they are optimal with respect to a specific objective function. To implement generate-and-search, one needs to define the search space. This is usually given as a list of the n variables (with given domains) and the space is defined as the Cartesian product of the domains of the variables.

Various search algorithms can be used in such a generate-and-search paradigm. However, due to rather high computational complexity of such problems, the standard exhaustive algorithms are not useful. Instead, heuristic search algorithms needs to be used. A typical example here are genetic algorithms (Goldberg 1989; Holland 1992; Rondeau 2007).

For example, in Rondeau's wireless system genetic algorithm (WSGA) approach (Rondeau 2007), the input data is a list of tunable knobs and values for each so that the values fall within the range of the data type for each knob. He used an XML file and a DTD file to represent this information. The XML file was used to provide the bounds, the step size and the number of bits for each knob, whereas the DTD file provided the minimum representation of the waveform to construct the chromosomes of the genetic algorithm. The information provided by the XML and DTD files is used to generate the chromosomes in the search space. The size of the search space can be reduced by various techniques, e.g. mapping to a feature space or using heuristics to cut some branches. Then the search algorithm will try to find the solution in the search space.

In contrast to data, *knowledge* is a useful representation of information that can be used to interpret the information (Rondeau 2007). For example, a sensor can collect the time and location information and provide them to the cognitive radio. This information is useless to the radio unless the radio knows what that information means about the potential use pattern, e.g. area of outage or high interference at a regular time of the day. Knowledge is used in *logical reasoning* for generating new knowledge from existing knowledge, whereas information does not have such an capability.

A knowledge-rich system, such as expert system, requires the creation of a "knowledge base" that captures the domain knowledge and represents it in a formal way so that the knowledge is processable by the machine. Typically, such knowledge is represented as rules, as discussed at the beginning of this chapter. Then the search algorithm (embedded in the inference engine) will search the knowledge base and try to find a solution. This search is directed by the knowledge captured in the rules and thus is more directed than e.g., the genetic algorithm search.

The success of a knowledge-rich approach depends on how good is the human's knowledge of the problem and whether it has been captured and represented in the knowledge base. If the problem-specific knowledge is abundant, then the knowledge-rich approach is likely to perform well. The most successful expert systems applications usually result from the fact that the human approach to solving the problem is already well understood by domain experts and stabilized (Mitola 2000).

The most distinguished benefit of the knowledge-rich approach is its capability to consider relations among the various input data. Information about relations is stored in the knowledge base. Thus for instance the reasoning process can combine data collected at different time and in different places and infer the the facts that are implicit in the data. These implicit facts can then be used to aid the decision making. Take the link adaptation problem as an example. The input of the the system is the knobs and meters from the $n - 1$ th transmission: $\{PowdB_{n-1}, trainPeriod_{n-1}, m_{n-1}, v_{n-1}, M_{n-1}, N1_{n-1}, N2_{n-1}, Memory_{n-1}, mSNR_{n-1}\}$. The knowledge-less approach only takes these parameters as unrelated input data and is not able to infer any additional facts that are relevant to the radio operation. On the other hand, the knowledge-rich approach can infer the implicit relationships among these parameters and the state of the radio, and then use this information to help the adaptation process.

When the search space is small, the knowledge-less approach can either search the space exhaustively, or use a problem specific algorithm that includes some IF/THEN statements to find the solutions. In order to be robust, the IF parts of the algorithm would have to cover the whole space and the THEN parts would have to provide information on how to process each of the data items from the search space that match a given IF part. However, the search spaces for various applications are rather large. For instance, in the CDMA system, there can be about 3,000 tunable knobs. The search space then is the Cartesian product of the domains of these knobs, i.e., the size of this space would be $3{,}000^n$. It is not feasible to write thousands of IF/THEN rules for each combination of the parameters from this space.

The knowledge-rich approach is more preferable in such cases. The knowledge base for this domain might have various classifications based upon the relationships in the KB. Each class thus would capture a large subset from the search space. Rules of the KB would then be written that would be applicable not to any specific point in the search space, but rather to the whole classes.

And finally, perhaps the greatest advantage of the knowledge-rich approach is its flexibility with respect to modifications. Whenever the need for modification arrives, only the knowledge (the rules) need to be changed; the inference engine remains the same for as long as the representation language has not been changed. This is in contrast with the knoledge-less approach in which any modification of the rules requires re-coding, re-compiling and so on.

Despite the many benefits described above, it is important to point out one acknowledged shortcoming of the knowledge-rich approach—new rules must be added by domain experts as the problem set evolves, and hence the knowledge-engineering bottleneck exists (Mitola 2000). Also, the knowledge-rich approach requires a large storage capacity, which may be a serious issue for a small-size radio.

7.2 Language Selection for Ontology-Based Radio

To represent knowledge we need a knowledge representation language. Such a language would be used to express both the data and the knowledge base. The knowledge base and the data would be used by the inference engine to infer facts. Moreover, the data and the knowledge represented in the KB could be communicated to other communicating nodes (cognitive radios). In the context of cognitive radio, the language (1) should be able to represent the knowledge of the cognitive radio domain and (2) should be usable in controlling the radio behavior to realize the requirements from different actors, e.g. consumers, first responders, service providers, manufacturers, lawmakers, etc.

There is agreement that a language must be accreditable, unambiguous, extensible, and interoperable. Currently, there are different IT communities working on developing such a language, e.g. IEEE 1900.5, E2R, and Wireless Innovation Forum MLM working group. Figure 7.1 shows a conceptual view of where standardized languages may play a role in the communication among various actors (Fette et al. 2008). The actors are shown at the outside of the figure. These are the individuals and organizations that are interested in communicating with regard to many issues. Examples of such issues are shown in the ovals, e.g. HW/SW portability, channel frequency modulations, etc. The intermediate layer shows some languages that the actors could possibly use. The Wireless Innovation Forum is working on a formal

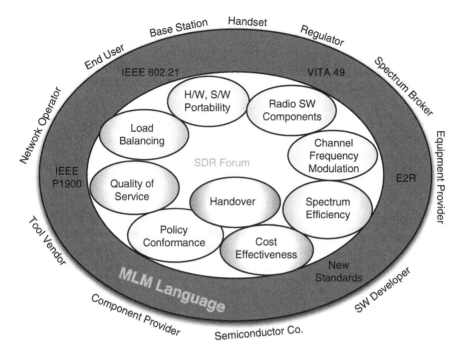

Fig. 7.1 Actors, issues and standard languages: a conceptual view (*source*: Fette et al. 2008)

language with computer processable semantics that could be used as a common
language among the various automated processes used by the actors to address their
communications and networking needs.

Dynamic Spectrum Access (DSA) is one of the issues of interest to regulators
as well as end users, who in this case might be represented by the software agents
running on various mobile devices. A number of emerging approaches to DSA systems employ rule-based mechanisms to adapt radio behaviors to application needs,
host system capabilities, in situ spectrum environment, and regulatory constraints.
The Defense Advanced Research Projects Agency (DARPA) neXt Generation (XG)
Communications Program proposed the use of non-procedural computer languages
and associated reasoners as a means for expressing and enforcing sets of policies
to enable and govern radio behaviors. Thus radios could roam the world while
autonomously enforcing spectrum access rules according to the policies provided by
the local spectrum governing authorities. Furthermore, that approach separates radio
technologies, regulatory policies, and optimizing techniques governing spectrum
access such that each of them could evolve asynchronously. Also, the End-to-End
Reconfigurability (E2R) project was working on a markup language (FDL, Function
Definition Language) for describing the functionality of various radio components
(Dolwin and Zhong 2007). More recently, the IEEE SCC 41 has begun efforts to
develop a set of interoperable and vendor- independent languages and architectures
for policy-based DSA systems (Kokar et al. 2008).

Specifically, we expect the language and associated semantic reasoning mechanisms to address the following areas as a minimum (Fette et al. 2008):

1. Capabilities of the nodes (e.g., frequency bands, modulations, MAC protocols, access authorizations, etiquette, bandwidths, and interconnections).
2. Networks available to a user (parameters, restrictions, costs).
3. Security/privacy (capability, constraints, policies).
4. Information types (an emergency call vs. just a "how are you" message).
5. Local spectrum situation (spectrum activity, propagation properties).
6. Network to subscriber and subscriber to network control (policies).
7. Manufacturer matters (hardware and software policy).
8. Types of users (authority, priority, etc.).
9. Types of data (Async., Isoc., narrow band, broad band, etc.).
10. Local regulatory framework (e.g., policies at a given geo location, time of day, emergency situation, etc.).
11. Time of day (at both ends of session and important points in between).
12. Geographic location (in three space, surrounding geography/architecture).

However, since cognitive radio is still under development, it is difficult to capture all the requirements for all the future needs, thus there is less consensus on the expressivity and computational modeling of such a language. In the following section, we will first talk about the distinction between imperative language and declarative language, and then present the available languages to express ontology and policy (Mitola 2009a).

7.2.1 Imperative Language Vs. Declarative Language

Basically, computer languages can be classified into imperative languages and declarative languages. Table 7.1 summarizes the differences between these two language types (Kokar and Lechowicz 2009).

In order to decide which language fits better our needs, we need to take a closer look at the requirements of cognitive radio.

As we have seen in the preceding chapters, cognitive radio must (1) be aware of the external communication environment and internal state and then (2) make decisions about the operating behavior to achieve the goal.

There are two concerns regarding the first aspect. First, the radio must not only "know" particular facts, but also be able to understand (or derive) the implications of the facts to its operation. For example, it is not sufficient for the radio to detect and record a dialog, but most importantly it must be capable of understanding the content of the dialog. Another example would be spectrum awareness. If the radio has detected an underutilized frequency, it must also know whether this frequency is assigned for public safety, analog TV or other usage. From this prospective, declarative languages can satisfy the awareness requirement. By combining an inference engine and a declarative knowledge base that relates various variables,

Table 7.1 Imperative language vs. declarative language

	Procedural/imperative language	Declarative language
Example	C, C++, Java	Prolog, SQL, OWL, SWRL
Algorithm	A sequential collection of operations/statement provides the algorithm	The algorithm is the inference engine. The input and output of the algorithm is provided by a collection of facts (clauses) and a goal defined by the user, e.g. a query is the input
Control structure	The control structure is partially determined by the ordering of the operations in the list and partially embedded in the control statement like if-then-else, do-while and do-until	The order of execution is determined in the way that the inference engine tries to find a solution to achieve the goal
"What" and "How"	A logic program consists of logic theory ("what") and deduction ("how"). The programmer needs to specify what needs to be done and how (in what sequence) can it be done	The programmer only needs to specify what needs to be done. The "how" part is accomplished in the inference engine
Modification	The whole program needs to be replaced because the program is the algorithm itself	Only the facts (clauses) and rules need to be replaced because they are only the input. The generic algorithm (the inference engine) is unchanged

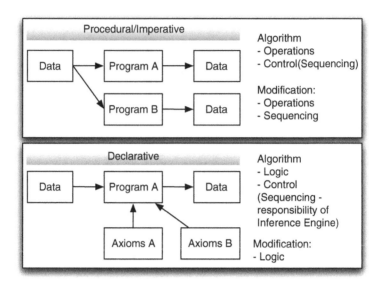

Fig. 7.2 Imperative language vs. declarative language

the radio can infer the implications of various operating states and environment conditions. Second, the radio is also required to be aware of its internal state. This is the feature called "reflection". The feature of reflection is part of most declarative languages and some procedural languages. For example, though a C++ program keeps values of its variables, it does not explicitly "know" its variables. To satisfy the internal awareness, a program must be able to query about the variables and reply to queries about its own variables, i.e., the program should be able to tell what the variables it has, the type and values of the variables (Fette 2009; Kokar and Lechowicz 2009).

At this point, we will discuss the second aspect—making decisions by the radio about its own behavior. The behavior of the radio is controlled by the algorithms implemented in the radio. If all the requirements and expected input data are known at the design time, then this information can be hard-coded into the program using an imperative language, which is more likely to provide better performance than using a declarative language. Unfortunately, the programmer is not always able to have all the information at the design time. An unexpected situation or goal may occur, which requires the radio to respond to surprise. In such a situation, using a declarative language is a better solution because all the knowledge in the knowledge base can be used to facilitate the inference engine's search for the best answer to an unexpected query, i.e., the query that was not programmed into the program. Conversely, due to the fixed control structure of the code, the procedural approach is not able to find an answer to an unexpected question. Furthermore, as shown in Fig. 7.2, in the declarative paradigm, if a program requires modification while it is running, we only need to change the logic rather than the algorithm itself (the algorithm is in the inference engine). On the other hand, in the procedural paradigm, the sequencing of the operations in the algorithm (the program) needs to be changed, which is much more difficult to achieve, compared to the declarative approach (Kokar and Lechowicz 2009).

For the above reasons, we come to the conclusion that a declarative language is a better fit for the language requirements of cognitive radio.

7.2.2 Ontology Language

The next question is—which declarative language to choose? There are different kinds of declarative languages. Cognitive radio requires that the language must be machine processable and understandable. In other words, the language must be a formal declarative language with formal syntax and semantics.

There are two kinds of knowledge that needs to be represented using such a formal declarative language: (1) the shared concepts between radios and networks; (2) the rules and policies that are used to control the behavior of the radio.

The shared concepts between radios and network are defined in common ontologies. In philosophy, ontology is the study of the nature of being or existence. The concept of ontology can be further extended to artificial intelligent, computer science and information science. Generally, it refers to a formal, explicit specification of a set of concepts in a specific domain and the relationships between these concepts. The term 'formal' means that the ontology is machine processable for the purpose of knowledge reuse and sharing (Studer et al. 1998).

In the cognitive radio domain, whenever a transmission is requested, there are at least three things that need to be expressed in ontology: (1) the capabilities of the radio, (2) the current environment of the radio and (3) the characteristics of the requested transmission.

Since different domains use different vocabularies, the use of ontology makes it possible to exchange information between radio agents across different organizations, providing a shared understanding of common domain. For example, by sharing a common ontology, the system strategy engine (SSE) maker, policy engine implementer, and regulatory policy author can consistently and unambiguously refer to the radio parameters and the relevant properties of the current radio environment such as frequency, power, location and signal characteristics (Mitola 2009a). In some cases, the adaptation is not only based on the local parameters but also on the parameters of the channel and other radios in the network. Hence, the use of ontology enables interoperability between radios and further facilitates multi-criteria adaptation on the network level.

Ontology can be classified into static facts and dynamic facts. Static facts, usually referred to as "T box", are the basic terms in a specific domain, usually including classes and properties. Dynamic facts, usually referred as "A box", are the facts only available as the radio is operating. They are usually the instances of the classes defined in T box.

The AI community has reached an agreement that a common language is needed to represent ontology. The most popular candidates include the Unified Modeling Language (UML) from the software engineering community and the Web Ontology Language (OWL) from the semantic web community. So far, the OWL has collected largest number of practitioners and supporters and the semantic web community is working on various ways to modify the expressivity of OWL. For the above reasons, we adopt OWL as the ontology representation language for cognitive radio. Though OWL has its limits on expressivity, there are other approaches to augment the expressivity of OWL, e.g. augment OWL with rules (Kokar and Lechowicz 2009).

7.2.3 Policy Language

In order to control and guide the behavior of a cognitive radio, an ontology language like OWL is not sufficient and thus needs to be extended. In particular, a language to express policies. Policies can be viewed as collections of rules. As described earlier in this Chapter, a rule consists of two parts—a hypothesis and a conclusion (also referred to as *body* and *head*, respectively. If the hypothesis part of a rule is satisfied, then the conclusion holds and is asserted. We refer to the collection of policies as the "R Box".

Policy engine is a software component that reasons with (or interprets) policies. The decisions of of the policy engine need to be *enforced* so that a particular communication device, or network of devices, obey a given set of policies during their operation.

Policy can be either external policy such as the permission to use frequency bands at specific location authorized by the FCC (stored in the Policy Conformance Reasoner) or internal policy for performance optimization (stored in the System Strategy Reasoner). The external policies are usually written by the regulator. The goal of regulatory policy is to specify the permissible transmission behavior of the radio, i.e., to describe conditions under which transmission is allowed. This kind of policy is not interested in the implementation details. Conversely, the internal policy usually is concerned with how to improve the performance of the radio. For example, a reduction of battery power may affect the ability to support multiple waveforms or to provide sufficient transmission power. An internal policy can be used to select or disable a low-priority waveform in order to save battery power and maintain support to the high-priority tasks.

In the policy-based approach, policies are separated from the implementation, which yields the benefits in the following aspects (Denker et al. 2009; Wilkins et al. 2007):

- The separation of policy and typical radio code enables the policy to be represented on a more abstract level and with an easier understandable semantics. In the current radios, policies are hard-coded into the radio and form an inseparable part of the radio's firmware. They are usually programmed using imperative language such as C, C++ and Java. These languages do not have an easily understood semantics and are not expressive enough to generally specify regulatory policies. Regulatory policies should be on a higher level than typical radio code and free from implementation details. As we've mentioned in the preceding section, policy is usually represented in a declarative language, which usually has an easier understood semantics and makes it easier to grasp the meaning of the policy.

- The policy-based approach decouples the definition, loading and enforcement of policy from device-specific implementations, which makes the certification process simpler and more efficient. The policy engine, policy and device can be accredited separately. The policy engine and each policy only need to be certified once and then loaded to any device without additional certification. A change to a component can be certified without accrediting the entire system. In this way, the cost of certification is shared across the network.
- The modification of the radio behavior becomes more flexible in a policy-based approach. For example, if a new policy is defined to adapt to a changing situation, the new policy can be dynamically loaded without recompiling any software on the radio.
- The policy-based approach can enable the policy and device to evolve independently, i.e., the radio technology can be developed in advanced of policies and vice versa.

Policy languages for cognitive radio have attracted interest in several radio and IT communities in the following aspects: (1) Spectrum Management (e.g. DARPA's XG and CoRaL radio policy languages), (2) Information Assurance and security, (3) Network Management (Strassner's DEN), and (4) Configuration Management (E2R and E3). However, there is no consensus on a common policy language so far (Mitola 2009a).

For our experiments, we used BaseVISor as the inference engine (policy engine) because the BaseVISor policy language is relatively simple and suitable for small scale experiments. In the BaseVISor rule language, both heads and bodies are expressed as triples. The triple-based rules are added to the rule base and then compiled into a Rete network, generating the nodes of the Rete network. Running the Rete network causes the rules to fire and facts to be added to the fact base. A particular rule is triggered when the triple patterns in the body of the rule mach the facts found in the fact base. The head of one rule may feed the body of another rule. Hence, the behavior can be flexibly controlled by the rules.

7.3 Structure of Ontology-Based Radio Reasoner

The architecture of a Cognitive Radio Node in Fig. 2.3 shows two reasoners—a PCR and a SSR. These reasoners use the rules in the R Box, the concepts in their T Boxes and the facts stored in their A Boxes as shown in Fig. 7.3. Moreover,

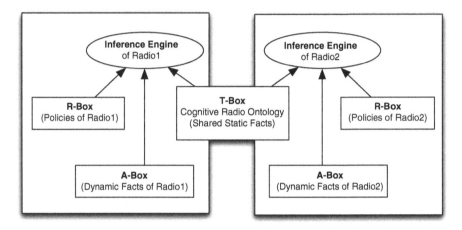

Fig. 7.3 Illustration of policy-based radio control

the collaborative adaptation scenarios discussed in this book include two Cognitive Radio Nodes (see Fig. 3.3). Each CR has an inference engine. As shown in Fig. 7.3, the two nodes share an ontology, but each has its own rule base and its own A Box.

Chapter 8
Cognitive Radio Ontology

8.1 Overview

In order to standardize the ontology-based approach to cognitive radio, a standard Cognitive Radio Ontology is needed. Towards this goal, we participated in the work of the Wireless Innovation Forum—the MLM (Modeling Language for Mobility) Work Group, whose goal was to come up with a standardized way of representing signaling among cognitive radios. With the help from the MLM WG, we developed a base ontology and submitted it as a contribution to the Forum. The CRO has been approved by the Wireless Innovation Forum as its recommendation (Li and Kokar 2010). It is expected that the CRO will provide opportunities for development of interoperable radios by independent vendors and lead to specifications/standards for data exchange to support the next generation capabilities.

The Cognitive Radio Ontology (CRO) includes:

- Core Ontology (covering basic terms of wireless communications from the PHY and MAC layers).
- Concepts needed to express the use cases developed by the MLM WG; only the use cases that relate to the PHY and MAC layers are included.
- Partial expression of the FM3TR waveform (structure and subcomponents, FSM).
- Partial expression of the Transceiver Facility APIs.

8.2 Principles of Modeling

8.2.1 Top-Level Classes

An upper ontology defines the most general concepts that are the same across different domains. Choosing an appropriate upper ontology as a reference model might be beneficial since this would help merging different ontologies into one

S. Li and M.M. Kokar, *Flexible Adaptation in Cognitive Radios*,
Analog Circuits and Signal Processing, DOI 10.1007/978-1-4614-0968-7_8,
© Springer Science+Business Media, LLC 2013

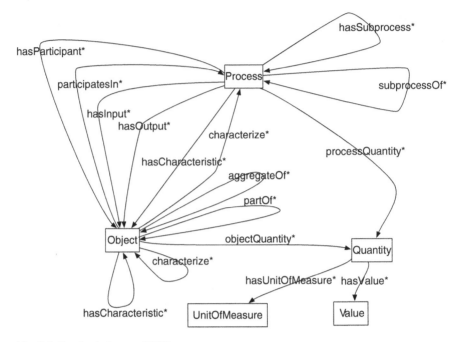

Fig. 8.1 Top-level classes of CRO

so that the common classes and properties (relations) are mapped correctly. In our work on CRO we did not fully incorporate an upper ontology, but rather patterned our ontology on DOLCE, the Descriptive Ontology for Linguistic and Cognitive Engineering (Masolo et al. 2009). DOLCE is based on the fundamental distinction between *Endurant*, *Perdurant* and *Quality*.

Endurant refers to the entity that is wholly presented at any given snapshot of time. Examples include material objects such as a piece of paper or an apple, and abstract objects such as an organization or a law. In our ontology we represent endurants by the class *Object*.

Perdurant is the entity that can be presented only partly at any snapshot of time. A process can have temporal parts and spatial parts. For example, the first movement of a symphony is a temporal part of a symphony, whereas the symphony performed by the left side of the orchestra is a spatial part of a symphony. In both cases, a part of a process is also a process itself. In our ontology perdurants are represented by the class *Process*.

We have identified the potential relationships between Object and Process as shown in Fig. 8.1.

1. An object cannot be a part of a process, but rather *participate* in a process. For example, a person is not a part of running, but rather participates in running.

2. The input and output of a process are objects. For instance, the input of modulation is a signal, where modulation is a process and signal is an object.
3. The capabilities of an object are a collection of processes. For example, a radio has the capabilities of transmitting and receiving. Here, a radio is an object; transmitting and receiving are processes.
4. The characteristics of an object or a process can be represented as objects. For instance, one of the characteristics of a transmitter is represented as *TxChannel-TransferFunction*.

Qualities in DOLCE are the basic attributes or properties that can be perceived or measured. Qualities cannot exist on their own; they must be associated with either an object or a process. All the qualities have values and some qualities have unit. The qualities without units are represented as data-type properties. The qualities with units are associated with a type of quantity.

In our ontology we introduce the class *Quantity*, which plays a similar role as Quality in DOLCE, but is closer to the engineering community language. Quantities can be subclassified into different types. For instance, a physical quantity represents a property of a physical object. Quantity carries three types of information: the type of the quantity (e.g., mass, length), the magnitude of the property (typically a real or integer number) and the unit of measurement associated with the given magnitude (e.g., [*kg*], [*m*]). In our ontology, quantity is a top-level class; it is further divided (sub-classified) into different types, such as length, frequency, time, etc. Each quantity is associated with a unit and a value.

Quantities are associated with objects or processes by the *objectQuantity* and *processQuantity* properties, respectively, as shown in Fig. 8.1.

There are two perspectives to representing the quantity property of an object or a process depending on whether the quantity has a unit or not. If the quantity has a unit then it is represented as a sub-property of *object-type* property *objectQuantity*. If it does not have a unit then it is represented as a *data-type* property.

For example:

1. *hasWeight* may be a property of Student; its unit is *kilogram*. Therefore, *hasWeight* is represented as a sub-property of object-type property *objectQuantity*. The domain of *hasWeight* is Student; the range is *Weight* (*Weight* is a sub-class of Quantity).
2. However, the property *studentID* for Student is represented as a data-type property. The domain of *studentID* is *Student*; the range is *Integer* (*Integer* is one of the built-in data types).

The same principle can be applied to represent a quantity property of a process. According to the classification described above, the top-level classes in our ontology are shown in Fig. 8.1; they include (1) Object, (2) Process, (3) Quantity, (4) Value, and (5) UnitOfMeasure.

Table 8.1 Examples of objects and process

Object		Process	
Alphabet	ChannelEncoder	State	ChannelCoding
AlphabetTableEntry	Detector	Transition	Detection
Channel	Modulator	Event	Modulation
ChannelModel	SourceEncoder	Action	SourceCoding
Component	Transceiver	AIS	Transceiving
Port	PNCode	Protocol	PNCodeGeneration
Agent	Packet	API	Multiplexing
Goal	PacketField	Method	... ...
DetectionEvidence	Network		
Signal	NetworkMembership		
Sample	Role		
Symbol	... ...		

8.2.2   Further Distinction: Object and Process

In this section, we are going to use some examples to further discuss the distinction between object and process. Table 8.1 shows an example list of objects and processes. All the examples are basic concepts within the cognitive radio domain.

8.2.2.1   Physical Object Vs. Non-physical Object

The distinction between physical object and non-physical object depends on whether an object has spatial properties (quantities). All the objects exist in time; but not all of them exist in space. The objects that exist in time and space, i.e., the ones with spatial location, are physical objects (Markosian 2000). Typically, the term physical object and material object are interchangeable. Conversely, non-physical objects only exist in time. For instance, signal is a physical object because it can be measured through time and space whether it is the signal conducted in the radio or the signal radiated in space.

Channel is the physical transmission medium; even though it cannot be visible by human eyes, it does indeed exist in both time and space and thus is a physical object. *ChannelModel* is a mathematical model that represents the characteristics of the channel. Most abstract mathematical concepts, such as equations and functions, are non-physical objects. *Goal* is the objective that an object intends to achieve. *Role* refers to what position a network member has in the network, e.g., master, slave or peer. Both goal and role are non-physical objects. *Detector* can refer to either physical object or non-physical object depending on what detector refers to. If detector refers to a physical device, e.g., a GPS as a location detector, then detector is classified as a physical object. This physical detector is visible and tangible; it has height and mass that represent its spatial properties. However, detector may also refer to the software module that performs the detection functionalities. In this case, detector is a non-physical object. The same methodology can apply to the analysis of some other objects listed in Table 8.1. For some concepts, if a precise definition is not given, then it is difficult to say whether it is a physical object or a non-physical object.

8.2.2.2 Object Vs. Process

Example 1 (Alphabet, Modulator, Modulation). The relationships among *Alphabet*, *Modulator* and *Modulation* are good examples to show the relationship between Object and Process.

Modulation is a process of varying one or more properties of a high frequency periodic waveform, called the carrier signal, with respect to a modulating signal. It usually takes digital signal as input and converts it to analog signal. Then the analog signal will be transmitted to the wireless channel. The changes in the carrier signal are chosen from a finite number of alternative symbols, which is called *alphabet*.

Alphabet, also known as *modulation alphabet*, is often represented on a constellation diagram. A constellation diagram represents the possible symbols that may be selected by a given modulation scheme as points in the complex plane. The coordinates of a point on the constellation diagram are the symbol values. If the alphabet consists of $M = 2^N$ alternative symbols, then each symbol represents a message consisting of bits. The index of each symbol value implies the bit pattern for that symbol. In real applications, *alphabet* is actually a lookup table that has the index and symbol value for each symbol. Regardless whether *alphabet* is a lookup table or a collection of symbols, *alphabet* presents itself as a whole at any snapshot of time; *alphabet* is a non-physical object.

Modulator refers to either an electronic device or a software module that performs modulation. In the former case, *modulator* is a physical object with input ports and output ports; in the latter case, *modulator* is a non-physical object that encapsulates a set of related functions, data and interfaces.

Example 2 (Specification vs. Implementation). Air Interface Specification (AIS) refers to "the set of transformations and protocols applied to information that is transmitted over a channel and the corresponding set of transformations and protocols that convert received signals back to their information content" (DYSPAN P1900.1 Working Group 2008). Typically, the specification of AIS is a document that establishes uniform criteria, methods, processes, etc. Therefore, such a specification is a non-physical object. In the DOLCE taxonomy, the specification of AIS can be further classified as a non-agentive social object, which is a subclass of non-physical object.

If two radios want to communicate with each other, both of them should implement the processes and methods in the AIS specification, though the details of the implementation may vary. Therefore, such an implementation of AIS is a process. In our ontology we have *AIS*, which is a subclass of *Process*.

Application Programming Interface (API) is a similar concept; it refers to an abstraction that a software entity provides of itself to the outside in order to enable interaction with other software entities. Typically, API contains the abstract description of a set of classes and functions. The software that provides the functions described by an API is said to be an implementation of the API. Therefore, it can be said that the specification of API is an object, whereas the implementation of the API is a process.

We use AIS and API as examples to demonstrate the difference between specification and implementation because they have something in common. Both of them are interfaces that provide a "standardization" to enable interaction between two objects. This standardization is an agreement that both of the objects must satisfy.

In general, we consider the specification of such an interface as an object whereas the implementation of this interface as a process.

In our ontology, we have both API and AIS categorized as subclasses of process. It is assumed that the terms API and AIS refer to the implementation, though the naming may not reflect this assumption.

8.2.3   Part–Whole Relationship

8.2.3.1   Aggregation Vs. Composition

In UML (the Unified Modeling Language), *aggregation* and *composition* are two different types of *association*; both of them represent a part–whole relationship. There is a distinction between aggregation and composition. *Aggregation* refers to the association relationship between two classes when a class is a collection or container of another class, but the contained class does not have a strong life cycle dependency on the container, i.e. when the container class is destroyed, its contents are not (Object Mangement Group 2007; Ambler 2004). For instance, *AIS* consists of one or more *protocols* for each layer that perform the layer's functionality. When the *AIS* no longer exists, its contained *protocols* are still there. Therefore, *AIS* is an aggregation of *protocols*. Conversely, *composition* has a stronger life cycle dependency between the container class and the contained class. When the container class is destroyed, its contents are destroyed, too. For instance, an *alphabet* table has several *alphabetTableEntry*, each *alphabetTableEntry* refers to a row in the table. When the *alphabet* table is destroyed, all the rows in that table no longer exist.

In UML 2, properties (associations) are formalized in the UML meta-model using the meta-classes *Association*, *Property*, *Class* and *DataType*. *Association* is an aggregation of two or more *properties*—instances of the *Property* meta class. One of the properties is linked with a class representing the domain of the association. Depending on whether the association range are objects or data types, the other property is linked with either a class or an instance of the *DataType* meta classes.

Fig. 8.2 Naming schemes for aggregation and composition

Preamble	Destination Address	Source Address	Control Field	... ...

Fig. 8.3 Packet frame structure

UML uses the *isComposite* Boolean-type meta-property of the *Property* meta-class to specify that a given aggregation is composite (strong aggregation). Since it is not possible to represent the property of a property in OWL, we use different naming schemes to distinguish between *aggregation* and *composition*. All the aggregation properties start with "aggregateOf" followed by the name of the range class. All the composition properties start with "compositeOf" followed by the name of the range class. Figure 8.2 shows an example to illustrate this.

8.2.3.2 Ordering of the Contained Entities

An instance of a class may contain an ordered collection of instances of other classes. The order of the contained instances must be explicitly represented. For instance, *Packet* is a composite of a sequence of *PacketField*. The ordering of the instances of *PacketField* is defined in the protocol. In this ontology, we use property *append* to represent the ordering of the contained instances, i.e. an instance can be appended to another instance. For instance, in the packet frame structure shown in Fig. 8.3, *Preamble*, *DestinationAddress*, *SourceAddress* and *ControlField* are instances of *PacketField* class. In the ontology, this ordering is specified by the *appendPacketField* property–a subproperty of *append*. A graphical view of the ontology capturing this structure can be found in Fig. B.4 in Sect. B.1.4 (Appendix B).

8.2.3.3 Examples of Part–Whole Relationship

The following examples will further illustrate how to represent part–whole relationships in this ontology.

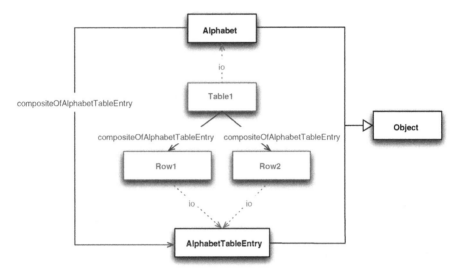

Fig. 8.4 Example of part–whole relationship (1): Alphabet and AlphabetTableEntry

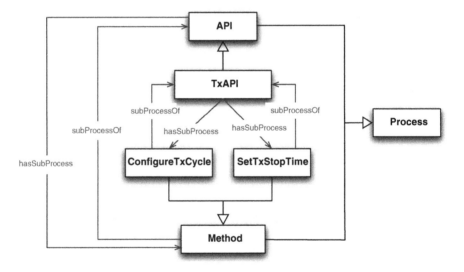

Fig. 8.5 Example of part–whole relationship (2): API and method

Example 1 (Alphabet and AlphabetTableEntry). In the example shown in Fig. 8.4, *Alphabet* is a composite of *AlphabetTableEntry*. Both of them are subclasses of *Object*. Table 8.1 is an instance of the class *Alphabet*. *Row1* and *Row2* are instances of the class *AlphabetTableEntry*. The "instance of" relation is shown with a dashed line.

Example 2 (API and Method). In the example shown in Fig. 8.5, instead of using "aggregateOf" or "compositeOf" that are used to represent the part–whole

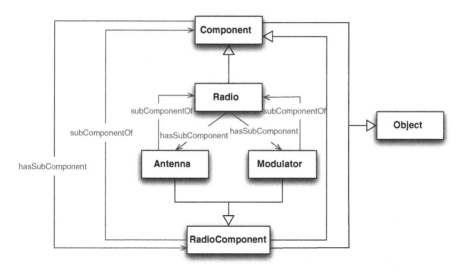

Fig. 8.6 Example of part–whole relationship (3): Radio and RadioComponent

relationship between objects, we use *hasSubProcess* to represent the part–whole relationship between processes. In general, a process can have other processes as its sub-processes. In other words, a process can be a sub-process of another process. For instance, an API contains the abstract definitions of a set of methods. Both *API* and *Method* are considered as *Process*. An *API* has several *Methods* as its sub-processes.

Example 3 (Radio and RadioComponent). The example shown in Fig. 8.6 is used to show the part–whole relationship between objects. A radio component consists of several sub-components, such as antenna and modulator. A pair of symmetric properties, *hasSubComponent* and *subComponentOf*, are used to represent the relationship between them. These two properties are sub-properties of the more general property *aggregateOf.* Although we could use *aggregateOf* to capture the part–whole relationship between components, we would then loose the more specific information about this relationship and, consequently, we would not be able to infer that say A and B are components from the information that they are related via the *subComponentOf* property.

Example 4 (Signal). In some cases, the *aggregateOf* relationship may need to be specialized to capture some specific aspects. For instance, in some cases an aggregate of two things may lead to a different class membership of the aggregate even though the particular components come from the same class. For instance, in Example 1, *AlphabetTableEntry* is part of *Alphabet,* but it is not an *Alphabet.* In Example 2, *Method* is part of *API,* but it is not an *API.* And in Example 3, *RadioComponent* is part of *Radio,* but it is not a *Radio.* However, both *RadioComponent* and *Radio* are subclasses of *Component.*

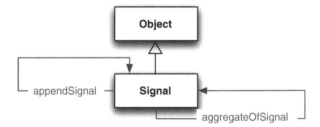

Fig. 8.7 Example of part–whole relationship (4): Signal

On the other hand, as shown in Fig. 8.7, an aggregate of two signals is also a *Signal*. However, the aggregation of signals must satisfy their temporal ordering. For this reason, the *aggregateOf* relation is complemented by adding another property *appendSignal*. The *append* keyword is used in this ontology to indicate the ordering of other objects.

8.2.4 Attribute, Property, Parameter and Argument

8.2.4.1 Attribute Vs. Property

There is a need for distinguishing between "property" and "attribute" (cf. the discussion in Mizoguchi n.d.). An *attribute* is a feature of an object that is independent of the context that the object is in. For instance, the *size* of a cup is this cup's attribute. Conversely, the property of an object depends on the context, for example, whether the cup is full or empty depends on the context, i.e., on the amount of liquid in the cup. So the "fullness" of a cup depends not just on the cup, but also on the liquid. In other words, this is a relation between the cup and the liquid in the cup. Relations in OWL are modeled as *OWL properties*.

The ontology presented in this paper is formalized in OWL. OWL, however, does not provide any simple means for an explicit distinction between attribute and property in the sense explained above. Take *PacketField* as an example. A packet consists of a sequence of packet fields. The size of a packet field is an attribute of *PacketField*, but whether a *PacketField* is optional or mandatory is a property since it depends on the context in which a specific packet field is used.

OWL only provides two types of properties: (1) object-type property, which links an individual to another individual, and (2) data-type property, which links an individual to an XML Schema data-type value (e.g., Integer, Boolean, etc.). Since we only use the features provided by OWL, both *packetFieldSize* and *isOptional* need to be modeled as data-type properties, i.e. *packetFieldSize* is linked to an integer value whereas *isOptional* is linked to a Boolean value (Li et al. 2008). So in summary, both attributes and properties are represented as either OWL object-type properties or OWL data-type properties.

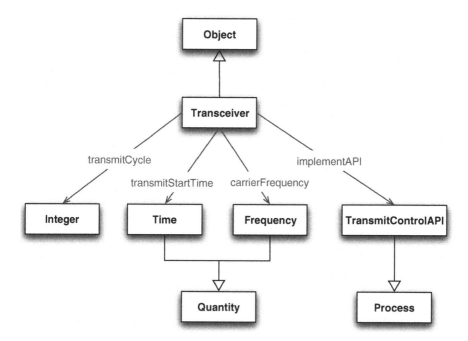

Fig. 8.8 Representation of properties and attributes: example of transceiver

8.2.4.2 Parameter and Argument

The concepts of "parameter" and "argument" are closely related. In mathematics, an *argument* is an independent variable and a *parameter* is a function coefficient. For instance, in equation $aX + bY = c$, variables X and Y are *arguments* whereas the function coefficients a, b and c are *parameters*. In computer science, *parameter* and *argument* are interchangeable. In engineering, properties of a system are the same as the attributes of a system. *Parameter* usually refers to combination of properties that is sufficient to describe a response of a dynamical system.

In our ontology, both *parameter* and *argument* are represented as either data-type or object-type properties, depending on the context.

8.2.4.3 Example: Attributes and Properties of Transceiver Subsystem

Figure 8.8 shows an example of how to represent the properties and attributes of the Transceiver subsystem. In this figure, *transmitCycle*, *transmitStartTime*, and *carrierFrequency* are the properties of *Transceiver*. *implementAPI* is an object-type property that shows the relationship between *component* and *API*.

TransmitCycle is the integer identifier that shall be set up during the creation to a specific value, which is then incremented by one for each newly created instance of the domain of the *transmitCycle* property. Since it is an integer number without any unit, it is modeled as a data-type property. The domain of this property is *Transceiver*; the range is *Integer*, which is a built-in data type.

TransmitStartTime refers to the transmit start time of the corresponding transmit cycle. Since it has the unit of second, it is represented as an object-type property. The domain is *Transceiver*; the range is *Time*, which is a subclass of *Quantity*.

The way of modeling *carrierFrequency* is similar as *transmitStartTime*. It is also represented as an object-type property. The domain is *Transceiver*; the range is *Frequency*.

In general, all the properties or attributes without unit of measure will be modeled as data-type properties, the range being one of the built-in data types, e.g., *Integer, Boolean, String*. Any properties or attributes with a unit of measure will be modeled as object-type properties whose ranges will subclasses of *Quantity*, e.g., *Time, Frequency, Location*. Each subclass of *Quantity* has a *Value* and a *UnitOfMeasure*. In this way, we can specify the values for those properties or attributes that have units.

8.3 Summary

In summary, the CRO has 230 classes and 188 properties, covering the basic terms of wireless communications from the PHY layer, MAC layer and network layer. The detailed description of CRO is shown in Appendix B.

Chapter 9
Implementation of Collaborative Link Optimization

In this chapter we describe the implementation of the architecture of Ontology Based Radio described in Chap. 2 and shown in Fig. 2.3. The main purpose of this implementation was to demonstrate the feasibility of the concept of collaborative flexible link adaptation discussed in this book. Thus first we need to show the platform that we had available and then how the architecture of this platform maps to the OBR architecture discussed in Chap. 2. In the rest of this chapter we describe the particular components of the architecture.

9.1 Implementation Platform

In order to implement the functionality of collaborative link adaptation the following capabilities were required:

1. Interaction between two OBRs that can facilitate the exchange of radio control messages (signaling).
2. Policies to control the radio behavior.
3. Inference engine to derive control decisions and formulate control messages.
4. Run time access (both read and write) to the variables of the radio.

To satisfy these requirements we used the Cognitive Radio Framework developed by Moskal et al. (2010). A top level view of this framework is shown in Fig. 9.1. This implementation covers most of the functionality of the OBR shown in Fig. 2.3, but not all. In particular, this implementation does not include any sensor interface since we were not interested in measuring the spectrum. Moreover, it does not include a Policy Conformance Reasoner; again since this was not needed for this case.

At the base of this implementation is a *Radio Platform* that provides the digital signal processing and software control, as well as the interfaces to communicate with the RF on the one side and with information source/sink on the other.

S. Li and M.M. Kokar, *Flexible Adaptation in Cognitive Radios*,
Analog Circuits and Signal Processing, DOI 10.1007/978-1-4614-0968-7_9,
© Springer Science+Business Media, LLC 2013

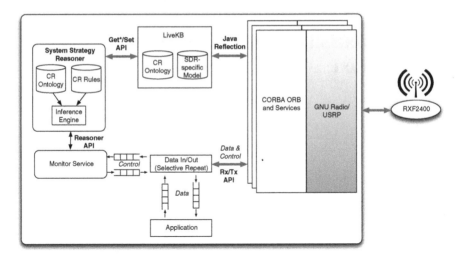

Fig. 9.1 Implementation architecture

Communication packages are filtered by the Data In/Out component which picks out control messages and sends them to the Monitor Service module. The content of communications packets are routed to the Application.

The *CORBA ORB and Service* is the middleware that enables the GNU Radio to act as a CORBA server and provide clients (upper layer applications) with means to transmit and receive data using the callback mechanism.

The *Data In/Out* module is responsible for distinguishing between control and data messages and passing the control messages to and from Monitor Service. All the incoming messages from the RF are first processed by the Radio Platform. Data messages are passed to the radio application (we call it Data Sink), whereas the control messages end up in the SSR. Similarly, all the outgoing control messages are generated by the SSR and then passed to a buffer. The data messages and control messages are merged in the buffer and then passed to the Radio Platform. After being processed in the Radio Platform, the messages are sent out through the RF channel.

Monitor Service (MS) implements the FIPA ACL protocols. The protocols are implemented by the state machines. It also invokes the System Strategy Reasoner component, collects it replies and then communicates with the other radio according to the FIPA ACL protocols.

More specifically, when a control message comes in, DI passes it to MS. MS unwraps the FIPA (see Sect. 9.4) part of the control message and passes the OWL encoded content to SSR. The content is written in OWL/RDF and thus can be processed by the inference engine. The outer part of the control message specifies the type of the control message and is defined using the FIPA ACL message structure.

The *LiveKB* component provides a generic GET/SET API, which allows the reasoner to access and adjust radio's parameters. The details of LiveKB are discussed in Moskal et al. (2010).

In the rest of this chapter we describe the particular functionalities of the implementation in more detail.

9.2 GNU/USRP

GNU Radio is a free software development toolkit that provides signal processing blocks to implement software radios using external RF hardware and commodity processors (Blossom 2004). The toolkit includes a library of modules written in C++. It also provides a Python interface that allows rapid application development by connecting the various blocks.

The Universal Software Radio Peripheral (USRP) is a high-speed USB-based board that enables general-purpose computers to function as software radios. In order to transmit and receive RF signals, the USRP motherboard is connected to daughter boards. The daughter boards are used to hold the RF receiver and transmitter. In our implementation, we used USRP1 as the motherboard and RXF2400 as the daughterboard. RXF2400 daughterboard is operated in the RF range from 2.3 to 2.9 GHz.

9.3 Policies for Link Adaptation

A policy is a set of rules written in a policy language. In our implementation, we use BaseVISor as the inference engine (Matheus et al. 2008). Policy rules are expressed in the BaseVISor syntax (BVR). The basic structure of BVR is RDF triple. An RDF triple consists of a *subject*, a *predicate* and an *object*. The subject and the object denote resources (things in the domain of discourse), and the predicate denotes a relationship between the subject and the object. BaseVISor is a forward-chaining rule engine optimized for handling facts in the form of RDF triples. The engine also supports XML Schema Data Types. Figure 9.2 is an example of a rule in the BVR form:

This rule states that if the SNR is smaller than 15 and larger than 10, then the performance is acceptable. In the BaseVISor syntax, the subject, predicate and object elements can be a resource, a XML data type or a variable. If an element is a resource, e.g., SignalDetector, then this element is defined in the ontology.

As is the case with some inference engines, it is possible to extend the BaseVISor functionality by adding new functions. These are called *procedural attachments* or *functions*. Figure 9.3 is an example of the invocation of a procedural attachment for computing an objective function.

```
<rule name="checkPerformance">
 <body>
   <triple>
    <subject variable="X" />
    <predicate resource="rdf:type"/>
    <object resource="rad:SignalDetector"/>
   </triple>
   <triple>
    <subject variable="X"/>
    <predicate resource="rad:signalToNoiseRatio"/>
    <object variable="SNR"/>
   </triple>
   <lessThan>
    <param variable="SNR"/>
    <param rdf:datatype="xsd:float">15</param>
   </lessThan>
   <greaterThan>
    <param variable="SNR"/>
    <param rdf:datatype="xsd:float">10</param>
   </greaterThan>
 </body>
 <head>
   <assert>
    <triple>
      <subject variable="X"/>
      <predicate resource="rad:performance"/>
      <object
rdf:datatype="xsd:string">acceptable</object>
    </triple>
   </assert>
 </head>
</rule>
```

Fig. 9.2 An example rule in BaseVISor format

Fig. 9.3 Example of
procedural attachment

```
<bind>
   <param variable="objFunc_PowdB"/>
   <computeObjFunc>
       <param variable="PowdB_new"/>
       <param variable="trainPeriod"/>
       <param variable="m"/>
       <param variable="v"/>
   </computeObjFunc>
</bind>
```

In this example, the objective function *<computeObjFunc>* is a user-defined procedural attachment. It has four arguments. It returns the value of the objective function and binds the value to variable *objFunc_PowdB*.

9.3.1   Policies for Link Establishment

The link adaptation is accomplished by executing the policy rules for which the pre-conditions are satisfied. Figure 9.4 shows an example of such executions (FIPA 2002g,i).

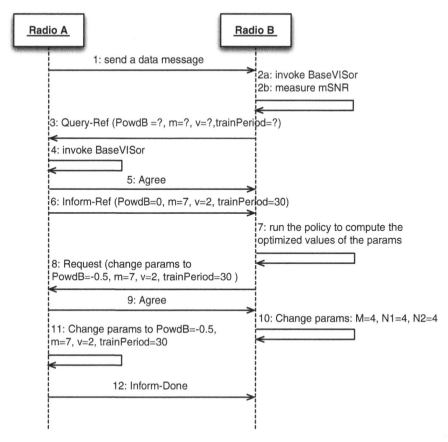

Fig. 9.4 Sequence diagram of link adaptation (1) : Query and Request

Suppose we initialize radio A as the transmitter and radio B as the receiver. After radio B receives a data message from radio A, radio B invokes its reasoner. Then a rule is fired to check the performance by measuring the $mSNR$. Radio B then first sends a "query" message to radio A, asking for the current values of its parameters. When radio A receives this query, it invokes its reasoner, which decides whether it can accept this query. If yes, then radio A sends an "agree" message to radio B, followed by an "inform-ref" message that contains the answer to the query. After radio B receives the answer, its policy rule is triggered to compute the new values of its local parameters and the parameters of radio A. Then radio B generates a request to radio A. The "request" message contains the new values of radio A's parameters. After radio A receives the request, it runs its reasoner to decide whether it can accept this request. If yes, then radio A sends an "agree" message to radio B and then sets its parameters accordingly. After radio A finishes setting its parameters, it sends an "inform-done" message to radio B. The following is an example of a rule (in pure text) for reacting to the received message:

Rule "checkPerformance":
If the radio receives a data message, then

1. Check mSNR
2. Generate a query message
3. Send the query to the originating radio

9.3.2 Policies for Link Adaptation

The goal of link adaptation is to minimize $objFunc$. However, the decrease of $objFunc$ will worsen the performance and decrease the $mSNR$. In other words, there is a tradeoff between the decrease of $objFunc$ and the improvement of $mSNR$.

We implemented three sets of policies with different preferences. Policy 1 decreases $objFunc$ as much as possible while not guaranteeing $mSNR$ within the acceptable range. Policy 2 decreases $objFunc$ to an intermediate level while maintaining $mSNR$ in the acceptable range, if possible. Policy 3 decreases $objFunc$ while guaranteeing $mSNR$ within the acceptable range.

The following shows the description of Policy 3; it contains four rules:

Rule 1:

If mSNR >15, then tune M, N1, N2 as follows:
    $M = -2, N1 = -2, N2 = -2$.

Rule 2:

If mSNR >12.5, then tune one of these parameters
    PowdB, trainPeriod, m or v as follows:

1. Compute the following:
     PowdB_new = min((12.5 − mSNR + PowdB), 0)
     trainPeriod_new = min(7.5 *(M+N1+N2), trainPeriod)
     m_new = 7
     v_new = max(v, floor((mSNR − 12.5)/6)+v)
2. Compute the following objective function values:
     objFunc(PowdB_new, trainPeriod, m, v)
     objFunc(PowdB, trainPeriod_new, m, v)
     objFunc(PowdB, trainPeriod, m_new, v)
     objFunc(PowdB, trainPeriod, m, v_new)

3. Choose the smallest objective function value from (2) and
   tune the corresponding parameter to the new value.

Rule 3:

If mSNR $<=$12.5, then tune one of these parameters:
   PowdB, trainPeriod, m or v.

1. Compute the following:
   PowdB_new $=$ min$((15 -$ mSNR$+$PowdB$)$, 0$)$
   trainPeriod_new $=$ min$(10 *($M$+$N1$+$N2$)$, trainPeriod$)$
   m_new $= 0$
   v_new $=$ max$($v, floor$(($mSNR$-15)/6)+$v$)$
2. Compute the following objective function values:
   objFunc(PowdB_new, trainPeriod, m, v)
   objFunc(PowdB, trainPeriod_new, m, v)
   objFunc(PowdB, trainPeriod, m_new, v)
   objFunc(PowdB, trainPeriod, m, v_new)
3. Choose the smallest objective function value from (2) and
   tune the corresponding parameter to the new value.

Rule 4:

If mSNR $<$10, then tune M, N1, N2 as follows:
   M $= +2$, N1 $= +2$, N2 $= +2$.
   To be interpretable by the reasoner, all the above policies were represented in the
BVR syntax.

9.4 Message Structure

As was mentioned earlier, in our implementation, we adopted the FIPA ACL
specifications (FIPA 2002f,g,i,c) to construct the control messages and design the
finite-state-machine of the MS component.

A control message contains two parts: (1) a set of message parameters, and (2) the
content. The message parameters provide information such as the type of message,
the sender and receiver, the conversation ID, etc. The content is described using a
FIPA ACL content language. The choosing of a content language depends on the
user's need. In our implementation, we chose OWL/RDF as the content language
because it has the machine interpretable syntax and can be directly processed by the
inference engine.

```
(REQUEST
        :sender(agent-identifier:name radioB)
        :receiver(set(agent-identifier:name radioA))
        :content
        "<?xml version=\"1.0\"encoding=\"utf-8\"?>
        <root>
            <triple>
                <subject resource=\"Run\"/>
                <predicate resource=\"FIPA\"/>
                <object resource=\"Request\"/>
            </triple>
            <rule name=\"request-from-radioA\">
                <body>
                    <triple>
                        <subject resource=\"Run\"/>
                        <predicate resource=\"FIPA\"/>
                        <object resource=\"Request\"/>
                    </triple>
                </body>
                <head>
                    <println>Changing tx_ampl to 0.309</println>
                    <set>
                        <param>/sdro:Radio/sdro:participatesIn/sdro
                :hasParticipant/sdro:txAmplitude</param>
                        <param datatype=\"xsd:float\">0.309</param>
                    </set>
                </head>
            </rule>
        </root>"
)
```

Fig. 9.5 Example of request message

```
(AGREE
            :sender(agent-identifier:name radioA)
            :receiver(set(agent-identifier:name radioB ))
            :reply-with radioB1295829968769
    )
```

Fig. 9.6 Example of agree message

The FIPA ACL specification defines 22 types of control messages (see Table 5.3). The sequence diagram shown in Fig. 9.4 is an example of two radios interacting with each other using some of these control messages.

Figure 9.5 is an example of a "request" message saying that radio B requests radio A to change its transmitter amplitude to 0.309.

If radio A gets the above request message and decides to accept this request, it sends an "agree" message back to radio B as shown in Fig. 9.6:

9.5   Message Exchange

As it was mentioned in Sect. 9.1, Monitor Service is responsible for processing the FIPA part of the control messages and then passing the OWL encoded content to the SSR. Since FIPA ACL specification provides only the protocols to support the message interactions between radios, a realization of these protocols had to be designed and implemented. For our implementation we used Moskal's implementation of these protocols (see Moskal et al. 2010 for details). In Moskal's design, the finite state machine of the Monitor Service was developed to implement the ACL protocols.

Recall the scenario shown in Fig. 9.4. At step 8, radio A receives a request from radio B and then runs its inference engine to decide whether or not to accept the request. If the request conflicts with the local regulations (e.g. the transmitter power is out of the permitted range), then radio A will send a "refuse" to radio B. However, the two radios still have the chance to negotiate with each other until an agreement is reached. According to the protocol provided by FIPA ACL, after radio B receives the "refuse" message, it can send a "call for proposal" to radio A. Then radio A can make a proposal to radio B. If the proposal is accepted, then radio A can change its parameters to the proposed values. Figure 9.7 shows the sequence diagram of the scenario (FIPA 2002f,g,i).

The finite-state-machine implemented in Monitor Service for the Call-For-Proposal (CFP) scenario is shown in Fig. 9.8. It corresponds to steps 10–15 in Fig. 9.9. Note that the "initiator" in Fig. 9.8 refers to radio B and the "participant" refers to radio A.

Figures 9.9 and 9.10 show the the finite-state-machine for the Query and Request shown in Fig. 9.4.

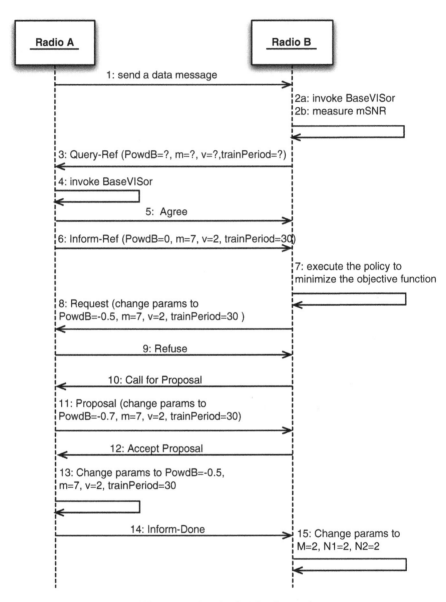

Fig. 9.7 Sequence diagram of link adaptation (2): Call-For-Proposal

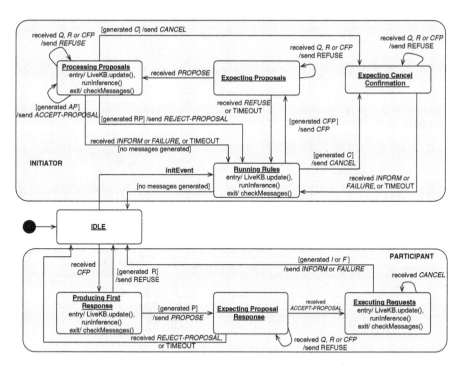

Fig. 9.8   Finite-state-machine of Call-For-Proposal (*source*: Moskal 2011)

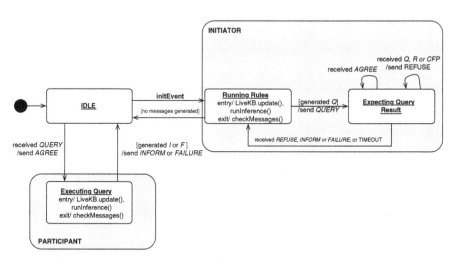

Fig. 9.9   Finite-state-machine of Query (*source*: Moskal 2011)

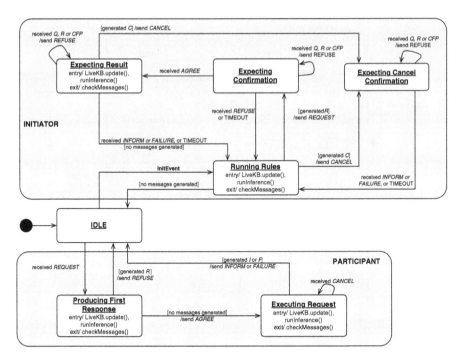

Fig. 9.10 Finite-state-machine of Request (*source*: Moskal 2011)

Chapter 10
Evaluations

The final step in the process of the development of a solution for collaborative link optimization was the experimentation with the developed architecture. Toward this aim, first, the developed link adaptation policies had to be verified, independently of the implemented architecture. We use MATLAB simulations for this purpose.

Having done this, the next step was to test the performance of the developed link adaptation approach from various perspectives. In particular, we focused on the improvement of performance due to adaptation, processing delays, communications overhead due to the exchange of control messages and the assessment of the inference capability of the inference engine.

10.1 Simulations of Policies in MATLAB

In order to evaluate the validity of the policies, we simulated the link adaptation scenario in MATLAB. We also used MATLAB to emulate a Rayleigh multipath channel between two radios. When a new control message arrives, the inference engine is invoked and policy rules are executed. The outputs of the rules include new values of the parameters for the next simulated transmission.

In order to evaluate whether the policies are able to adapt to the change of the channel environment, we linearly increase the number of multipath from 2 to 16. Assume the radios are operating in half-duplex mode. The default parameter values are: $PowdB = 0$, $m = 3$, $v = 1$, $trainPeriod = 100$, $M = 2$, $N1 = 2$, $N2 = 2$. First, we set the number of multipath to 2. Then radio A sends the 1st data message to radio B. When radio B receives the 1st data message, it measures $mSNR$ and $objFunc$. Based on the current values of $mSNR$ and $objFunc$, the two nodes follow the steps described in Fig. 9.4 to compute the parameters for the second data message and then set their parameters to the new values. Then we change the channel environment by setting the number of multipath to 4. After that, radio A sends the second data message to radio B and repeats the above steps. In total, radio

S. Li and M.M. Kokar, *Flexible Adaptation in Cognitive Radios*,
Analog Circuits and Signal Processing, DOI 10.1007/978-1-4614-0968-7_10,
© Springer Science+Business Media, LLC 2013

A sends eight data messages to radio B. The simulation results and the comparison of these three policies are shown in Fig. 10.1. It can be seen that without link adaptation, *objFunc* remains at the same value and *mSNR* fluctuates as the number of multipath changes. With link adaptation, the value of *objFunc* is significantly decreased. Policy 1 decreases *objFunc* by 66 % at the price of decreasing *mSNR* by 1.83 dB. Policy 2 decreases *objFunc* by 55 % at the price of decreasing *mSNR* by 0.83 dB. Policy 3 decreases *objFunc* by 36 % while increasing *mSNR* by 0.09 dB. So the policies clearly have an impact on the performance of the two radios and the results are in line with is reasonably expected from this kind of scenario.

10.2 Policy Execution Results

The results of policy execution on the implemented architecture for the scenario described below are shown in Fig. 10.2. In this scenario, radio A keeps sending an image to radio B. Radio B measures the *mSNR* and then initiates the link adaptation process. The horizontal axis at the bottom shows the number of packets received. The vertical axis on the left shows the value of the objective function. The vertical axis on the right shows the measured *mSNR* at radio B. Note that the objective function used in the implementation is the power efficiency, i.e., the information bit rate per transmitter watt of power (in $Gbit/watt \cdot sec$); it is the reciprocal of the objective function that is used in the MATLAB simulation. During the experiment, we moved the two radios around and deliberately changed the distance between them. It resulted in some changes of the channel environment and thus some fluctuation of the *mSNR*. It can be seen that when *mSNR* is too large, the radios adjust their parameters to lower the *mSNR* and thus increase the power efficiency. When *mSNR* is too small, the radios adjust their parameters to increase the *mSNR* by the price of lowering the power efficiency.

10.3 Evaluation

The goal of the link adaptation use case was to maximize the power efficiency subject to a set of constraints. In the case when there is no link adaptation, the values of the knobs of the transmitter and receiver are fixed, i.e., the radios keep using the initial values of their parameters during the transmission and there is no change of the parameters unless the users manually change them. Thus the power efficiency remains at the same level while the *mSNR* fluctuates as the channel environment changes. Take the transmitter power as an example. If the initial transmitter power is set to a high level, then the radio may waste energy when the channel environment is "good" and *mSNR* is very high. In such a case, the transmitter can use a lower transmitter power to increase the power efficiency yet still maintain the *mSNR* in an acceptable range. Conversely, if the initial transmitter power is set to a low

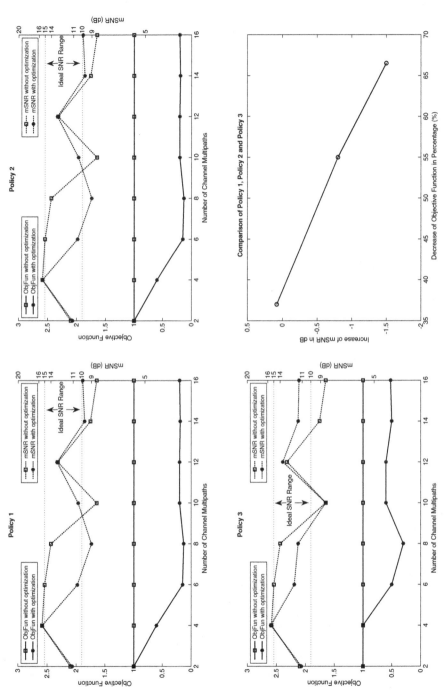

Fig. 10.1 MATLAB simulation results: comparison of Policy 1, Policy 2, Policy 3

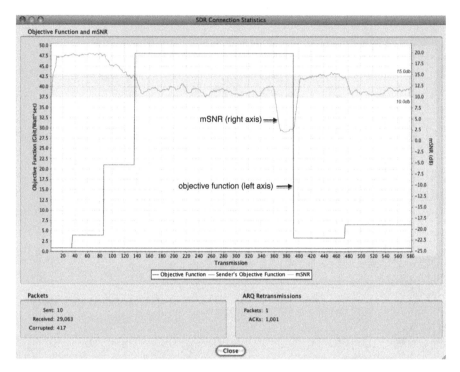

Fig. 10.2 Policy execution results

level, then it may lead to an increase of the number of lost packets or corrupted packets when the channel environment is "bad" and *mSNR* is very low. In such a case, the transmitter shall use a higher transmitter power in order to bring the *mSNR* back to an acceptable range. Thus, it is necessary to adapt the radio parameters to the change of the channel environment. This can be achieved by the approach described in Sect. 9.3.2. In order to evaluate the benefits and costs of the ontology and policy based radio adaptation, the rest of this section will assess the performance improvement, processing delay, control message overhead, inference capability and flexibility attributed to this approach.

10.3.1   Performance Improvement

In our experiments, radio A is the transmitter and radio B is the receiver, both of them are operating in half-duplex mode. In each run, radio A sends an image of 10,000 pixels to radio B. Each pixel is sent as an individual packet and the size of each packet is the same. Assume that the initial transmitter power is 15 dBm. In the case when there is no adaptation, radio A uses the same transmitter power

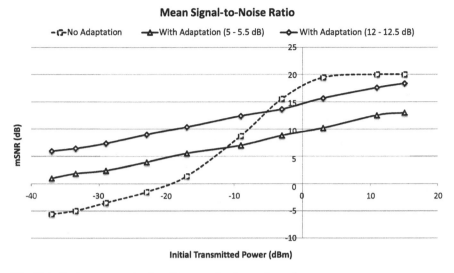

Fig. 10.3 Performance evaluation (1): mean signal-to-noise-ratio

to send all the 10,000 packets. In the case when there is adaptation, radio A will use the initial transmitter power to send the first few packets until the measured *mSNR* at radio B is out of range. The acceptable range of *mSNR* is specified in the adaptation policy. For comparison purpose, we developed two sets of adaptation policies. The first one specifies the acceptable *mSNR* range from 5 to 5.5 dB; the second one specifies the range from 12 to 12.5 dB. Then radio B will trigger the adaptation policy and compute a new value of the transmitter power, then it will request radio A to change its transmitter power accordingly. The power adaptation process continues until radio A finishes sending all the packets.

We change the initial transmitter power from −37 to 15 dBm in uneven increments (see Fig. 10.3). For each initial transmitter power, we run the experiments for 10 times for the case without adaptation and another 10 times for the case with adaptation. Then we compute the average power efficiency, mean signal-to-noise ratio and average corrupted packet rate for each case.

Figures 10.3–10.5 show the comparison results of the communications link with adaptation and without adaptation, in terms of mean signal-to-noise ratio, power efficiency and corrupted packet rate. All the x-axes are the initial transmitter power.

It can be seen that (1) when the initial transmitter power is smaller than −10 dB, the use of adaptation can yield smaller power efficiency, but the corrupted packet rate is smaller due to higher *mSNR*. Smaller corrupted packet rate means that there will be less traffic imposed to the network because the radios have less need to re-send the packets. (2) When the initial transmitter power is larger than −10 dB, the use of adaptation will increase the power efficiency, yet it will not increase the corrupted packet rate, i.e. in Fig. 10.5, when initial transmitter power is larger than −10 dB, the blue line ("with adaptation") and the red line ("no adaptation") are almost on the same level.

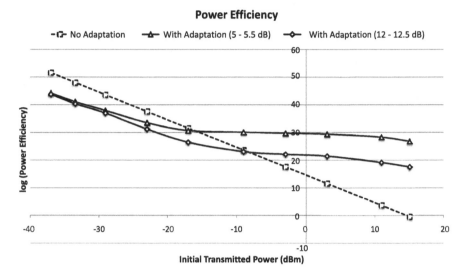

Fig. 10.4 Performance evaluation (2): power efficiency

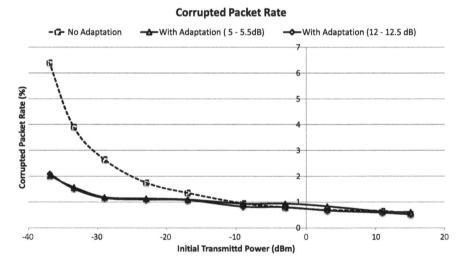

Fig. 10.5 Performance evaluation (3): corrupted packet rate

The comparison of the overall performance is shown in Fig. 10.6. It can be seen that if we want to constrain the corrupted rate at a certain level, then the use of link adaptation can yield higher power efficiency; in other words, if the power efficiency remains the same, then the use of link adaptation can decrease the overall corrupted rate.

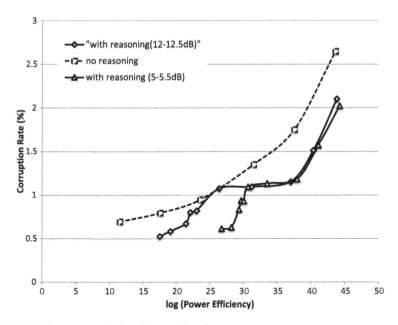

Fig. 10.6 Performance evaluation (4): overall performance

10.3.2 Processing Delay

As shown in Fig. 9.4, in our implementation of the link adaptation use case, the radio is able to generate five types of messages: Query-Ref, Agree, Inform-Ref, Request, and Inform-Done. To make the case simpler, we assume that radio A always agrees to an incoming query or request. To implement this, the MS will generate an "Agree" when it receives a "Query-Ref" or "Request", rather than passing it to the SSR and let the inference engine make the decision. All the other four types of control messages are generated by the inference engine.

In order to evaluate the processing delay imposed to the system due to the use of ontology and policy based approach, we measured the response time needed to generate each type of control message in the inference engine. For this purpose we used the time stamping facility of the MAC OS, version 10.5.8; processor 2.5 GHz Intel Core 2 Duo.

The response time depends on the type of control message and the size of the search space, i.e. the number of facts (triples) in the knowledge base. For the evaluation purpose, we created five ontologies with different sizes: each of which was used as the T-Box shown in Fig. 7.3. For instance, we use the ontology with 500 triples as the T-Box, then we run the sequence shown in Fig. 9.4 for 50 times and measure the average response time for each control message generated by the inference engine. Then we run the experiment again using the ontology with 1,000 triples, 1,500 triples, 2,000 triples and 2,500 triples. Figure 10.7 shows the average

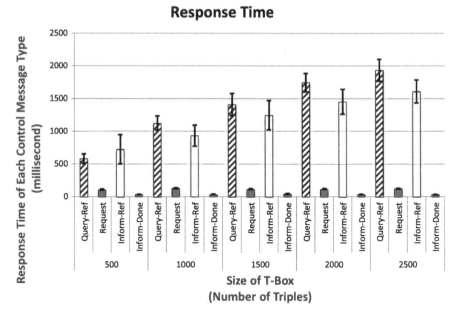

Fig. 10.7 Response time of each control message

response time of each control message type for T-Box with different size along with the standard error for each one. It can be seen that: (1) the response time to generate "Query-Ref" and "Inform-Ref" increases proportionally to the size of the T-Box. (2) The response time to generate "Request" and "Inform-Done" does not increase as the size of the T-Box increases and it is much less than the time to generate "Query-Ref" and "Inform-Ref".

10.3.3 *Control Message Overhead*

To evaluate how much reduction can be achieved, we created 17 different control messages encoded in OWL, each of which represents either a query, a request, or an inform of a parameter in the radio. Figure 10.8 shows the original size and the compressed size of each control message. Two compression tools are used: Gzip and Xmill. It can be seen that Gzip has a higher compression ratio than Xmill. In average, the compression ratio of Gzip is 0.4 and XMill is 0.5.

Though the text message format results in larger message size and more processing time, the advantage of this approach is that (1) XML file containing errors (due to communication) can still be processed without error messages. Note that an erroneous Gzip XML file can be decompressed without error messages, but an erroneous XMill XML will not be decompressed and it will display error messages; (2) the costs of updating XML file is less than VMF (Rhyne et al. 2002).

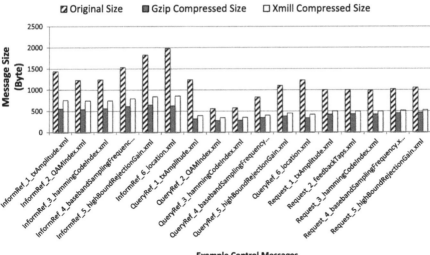

Fig. 10.8 Control message overhead

10.3.4 Inference Capability

One of the advantages of the OBR approach is its inference capability. In the case of link adaptation, the radios need to exchange information about their knobs and meters. So theoretically, radios might need to send values of 3,000 of such parameters. This would impose a lot of communications burden leading to a high need for spectrum. However, not all the 3,000 parameters are needed in the link optimization. Generally, in each transmission, these 3,000 parameters can be divided into three groups: (1) parameters that need to be changed; (2) parameters that are fixed; (3) parameters that we don't care. For example, in our use case, we only care about seven parameters: $PowdB$, $trainPeriod$, m, v, M, $N1$, $N2$. Suppose radio B needs to send a command to radio A, requesting it to change the values of $PowdB$ to -5.5 and keep $trainPeriod$, m, v unchanged. To address this scenario, we can extend the CRO to include the class *Configuration*, shown in Fig. 10.9. Each configuration class specifies the setting for a combination of tunable knobs. Potentially, each configuration may include combinations of settings for thousands of tunable knobs.

Assume that initially radio A and B share a common ontology as shown in Fig. 10.9. Since the two radios come from two different vendors, the ontology can be extended by each vendor by adding sub-classes to the common configuration ontology. For instance, Radio A has subclass *Config1* and *Config2*, while Radio B has subclass *Config3* and *Config4*. When the two radios request each other's knobs and meters, they would not be able to do this because they don't understand the

Fig. 10.9 Extend CRO with configuration class

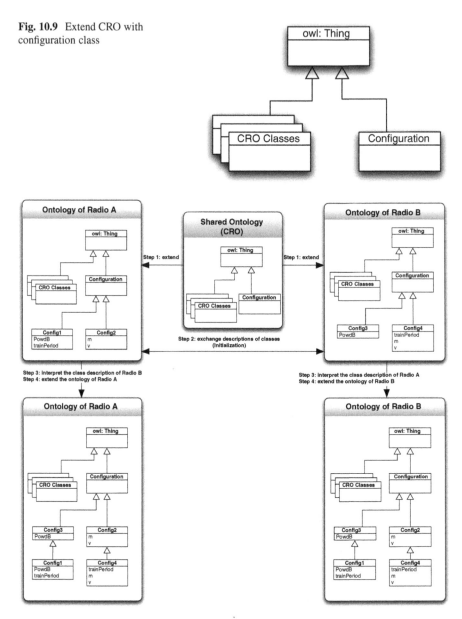

Fig. 10.10 Inference capability: a configuration example

requests due to the lack of the definitions of the additional classes. However, since both of them understand OWL, they can exchange the class descriptions as OWL expressions and extend their ontologies with the additional classes. After this, they can use their own OWL reasoners for inference over the extended ontologies. The whole process is shown in Fig. 10.10.

```
<Config3rdf:ID="Config1_instance">
  <PowdB>
    <Power rdf:ID="Power_instance">
      <hasValue>
        <FloatValue rdf:ID="FloatValue_instance">
          <hasFloat rdf:datatype="xsd:float"
          >-5.5</hasFloat>
        </FloatValue>
      </hasValue>
      <hasUnitOfMeasure>
        <UnitOfMeasure rdf:ID="dBm"/>
      </hasUnitOfMeasure>
    </Power>
  </PowdB>
</Config3>
```

Fig. 10.11 Instance of Config3

Then in order to request radio A to change the values of *PowdB* to −5.5 and keep *trainPeriod*, *m*, *v* unchanged, radio B only needs to send a command message to radio A containing the following information: (1) the name of class *Config4*; (2) an instance of class *Config3* that includes the new value of *PowdB*.

The OWL representation of the instance of class *Config3* is shown in Fig. 10.11.

When radio A gets this request, since it has the definition of this configuration and thus is able to understand the request and set the parameters appropriately. This simple example shows that if the radios have this kind of information encoded in their ontologies or rules, they do not need to send all the information, but instead may infer the rest of the values locally. In particular, in this example, radio A can infer what needs to be changed and what should stay unchanged.

In order to evaluate the inference capabilities of the ontology approach we created 50 different ontologies written in OWL with different sizes ranging from 500 to 2,700. The size of the ontology refers to the number of facts (or triples) in the T-Box of the knowledge base. Then we pass each ontology into the inference engine and let it to infer new facts and add such new facts to the knowledge base. In Fig. 10.12, the blue bar shows the number of initial facts in the knowledge base before doing any inference, the red bar shows the number of facts in the knowledge base after doing the inference. It can be seen that after doing the inference, a large number of new facts are added to the knowledge base.

Intuitively, if we use the XML to encode the facts in the knowledge base, then we must send all the information explicitly. If we use the ontology approach, then the radio only needs to send parts of the message, while the rest of the information can be inferred locally by the inference engine based on the generic knowledge encoded.

To further address the comparison between XML and OWL for the purpose of communication between two radios, assume the facts in the knowledge base (both explicitly represented and those that can be inferred) is the set $s = \{t_1, t_2, t_3 \ldots \ldots t_N\}$. Let X denote a fact to be chosen and sent to the other radio. Assuming that the choice is uniformly distributed, then the probability that t_i is chosen and sent is $P(X = t_i)$, which equals to $1/N$. Further assume that any given

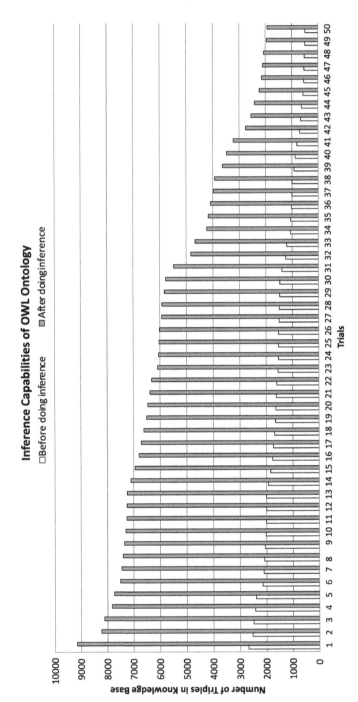

Fig. 10.12 Inference capability of OWL ontology

fact requires L bytes to represent in both XML and OWL. Since some of the facts, if represented in OWL, can be inferred locally, they don't need to be transmitted. Thus on the average, the bandwidth required to send OWL facts will be smaller than for sending XML facts.

Take the 50*th* trial in Fig. 10.12 as an example. Initially, there are 500 facts in the knowledge base. After doing the inference, 1,433 new facts are added to the knowledge base, therefore the ratio of the facts encoded in OWL that need to be transmitted to the facts encoded in XML that need to be transmitted is:

$$r = \frac{500}{500 + 1433} \approx 0.26.$$

The average r over the 50 trials in Fig. 10.12 is approximately 0.27. It can be seen that using the ontology approach, for each chosen fact, the transmitter only needs to send 27 % of the information, and then the receiver can infer the rest locally, resulting in a reduce of communication traffic imposed to the network.

Appendix A
Typical Knobs and Meters

The commonly used meters of the cognitive radio include (Cognitive Radio Working Group 2008):

Link quality measurements in the physical layer:

- Bit error rate (BER)
- Frame error rate (FER)
- Signal-to-noise-ratio (SNR)
- Received-signal-strength (RSS)
- Signal-to-noise-plus-interference-ratio (SINR)

Channel selectivity measurements in the physical layer:

- Time selectivity of channel (Doppler spread)
- Frequency selectivity of the channel (Delay spread)
- Space selectivity of the channel (Angle spread)
- LOS and NLOS measure of the channel

Radio channel parameters (including path loss, long and short term fading):

- Noise power
- Noise plus interference power
- Peak-to-average-power-ratio (PAPR)
- Error-vector-magnitude (EVM)
- Cyclostationary features

S. Li and M.M. Kokar, *Flexible Adaptation in Cognitive Radios*,
Analog Circuits and Signal Processing, DOI 10.1007/978-1-4614-0968-7,
© Springer Science+Business Media, LLC 2013

Link quality measurements in the MAC layer:

- Frame error rate (CRC check)
- ARQ request rate (for data communication)

Other possible measurements in the networking layer:

- Mean and peak packet delay (for data communication)
- Routing table or routing path change rate (for ad-hoc and sensor networks)
- Absolute and relative location of nodes (location awareness), velocity of nodes, direction of movement

The Cognitive Radio Working Group in the Wireless Innovation Forum provides an example list of operational parameters (knobs) that can be adapted and optimized (Cognitive Radio Working Group 2008):

Link and network adaptation:

Physical layer writable parameters

- Transmitted power
- Channel coding rate and type
- Modulation order
- Carrier frequency
- Cyclic prefix size (in OFDM based systems)
- FFT size, or number of carriers (in OFDM based systems)
- Number of pulses per bit (in impulse radio based ultra-wideband systems)
- Pulse-to-pulse interval, i.e. Duty cycle (in ultra-wideband systems)
- Antenna parameters in multi-antenna systems (such as antenna power, switching antenna elements, antenna selection and beam-forming coefficients, etc.)
- RF impairment compensation parameters, etc. (including many other system-specific and writable parameters)

Mac layer writable parameters

- Channel coding rate and type
- Packet size and type
- Interleaving length and type
- Channel/slot/code allocation
- Bandwidth (such as the number of slots, codes, carriers, and frequency bands, etc.)
- Carrier allocation in multi-carrier systems; band allocation in multi-band systems

Other writable parameters

- Cell assignment (in hierarchical cellular)
- Routing path/algorithm (for multi-hop networks)
- Source coding rate and type
- Scheduling algorithm
- Clustering parameters (for clustering based routing and network topology)

Related to context awareness:

- Service personalization (to adapt services to the context such as user preferences, user location, network and terminal capabilities)

Receiver adaptation:

- Channel estimation, synchronization, frequency offset parameters adaptation
- Soft information generation adaptation
- Equalization/demodulation parameters adaptation
- Interference/noise cancellation parameters adaptation
- Receiver antenna selection/combining adaptation
- Receiver filter adaptation

Constraints in employing adaptation:

- Constant BER (ensuring that the desired BER requirement is satisfied)
- Constant FER
- Maximizing the overall system throughput
- Minimizing the network power dissipation (especially critical for power efficient network design such as wireless sensors networks)
- Minimizing the average and peak delay
- Maximizing the system capacity
- Maximizing the user's perception of the video/speech quality or other services

Appendix B
Cognitive Radio Ontology

B.1 Object

B.1.1 Alphabet and AlphabetTableEntry

Alphabet is a lookup table that participates in the *Modulation* process. In digital modulation, an analog carrier signal is modulated by a digital bit stream. The changes in the carrier signal are chosen from a finite number of M alternative symbols, which is called *Modulation Alphabet* or *Alphabet*. Each row of the Alphabet table is an instance of *AlphabetTableEntry* class. Hence, *Alphabet* is an aggregation of *AlphabetTableEntry*. The *AlphabetTableEntry* class has two properties: (1) *hasIndex*, and (2) *hasSymbolValue*, as follows (Table B.1).

A modulation alphabet is often represented on a constellation diagram. A constellation diagram represents the possible symbols that may be selected by a given modulation scheme as points in the complex plane. The Real and Imaginary axes are often called the in-phase (I-axis), and the quadrature (Q-axis). The coordinates of a point on the constellation diagram are the *symbol values*.

If an alphabet consists of $M = 2^N$ alternative symbols, then each symbol represents a message consisting of N bits.

B.1.2 Channel

Channel refers to the physical transmission medium between the transmitter and the receiver. For example, in a cellular system, the transmission link between the Mobile Station (MS) and the Base Station (BS) can be divided into (1) Forward Channel (from BS to MS) and (2) Reverse Channel (from MS to BS). Typically, both Forward and Reverse Channel have sub-channels for the transmission of either data messages or control messages. For instance, in the IS-95 system, there are three types of

S. Li and M.M. Kokar, *Flexible Adaptation in Cognitive Radios*,
Analog Circuits and Signal Processing, DOI 10.1007/978-1-4614-0968-7,
© Springer Science+Business Media, LLC 2013

Table B.1 Example of
Alphabet Table

Index	Symbol value	Bit pattern (optional)
0	$1 + j$	00
1	$-1 + j$	01
2	$-1 - j$	11
3	$1 - j$	10

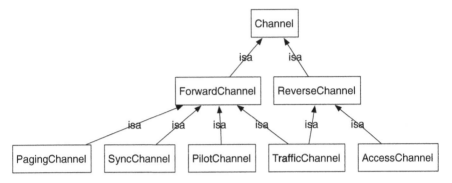

Fig. B.1 Subclasses of Channel

overhead channels in the forward link: pilot channel, synchronization channel and paging channel. The modulation, bandwidth, data rate and the multiplexing scheme of each channel are specified in the Air Interface Specification (AIS).

The subclasses of Channel are shown in Fig. B.1. The relationships among *Channel*, ChannelModel, Multiplexing and Modulation is shown in Fig. B.2.

B.1.3 ChannelModel

Each channel is associated with one or more than one channel models. *ChannelModel* is used to represent the estimated effects of the propagation environment on a radio signal. Well known channel models include Additive White Gaussian Noise (AGN) channel, Rayleigh fading channel and Rice fading channel. The subclasses of *ChannelModel* are shown in Fig. B.3.

B.1.4 Packet and Packet Field

A *Packet* is a formatted unit of data transmitted between radios.

First, a packet consists of a sequence of packet fields, thus it is an aggregation of *PacketField*.

Second, a packet field can be appended to another packet field, forming an ordered collection of packet fields.

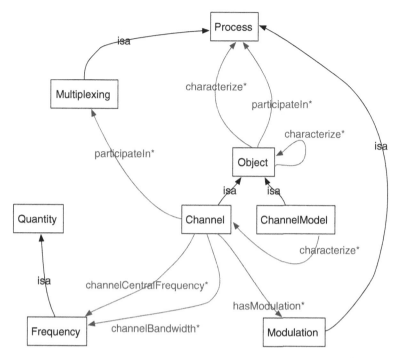

Fig. B.2 Relationships among Channel, ChannelModel, Multiplexing and Modulation

Fig. B.3 Subclasses of
ChannelModel

Third, the ordering of a *PacketField*, the size of each field, and the values allowed in each field are defined in a protocol. Therefore, both *PacketField* and *Packet* are associated with a protocol. The relationships among *Packet*, *PacketField* and *Protocol* are shown in Fig. B.4.

In the OSI or TCP/IP model, each layer has various protocols, each of which defines the syntax and the semantics of packets. Packets of different layers may have different names. For instance, a data link layer packet is usually called a *frame*. For this reason, *Packet* is divided into several subclasses for the lower three layers, shown in Fig. B.5, including *DLLFrame*, *NLPacket*, and *PHYPacket*.

PacketField can be of different types. Generally speaking, a packet should at least consist of *training sequence* and *payload*. In this ontology, we only capture these two types. The taxonomy of *PacketField* class is shown in Fig. B.6.

Fig. B.4 Relationships among Packet, Packefield and Protocol

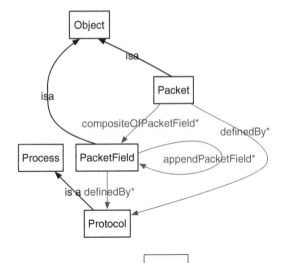

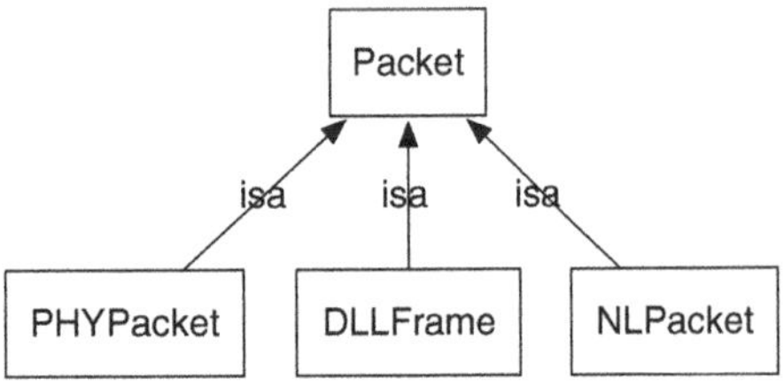

Fig. B.5 Subclasses of Packet

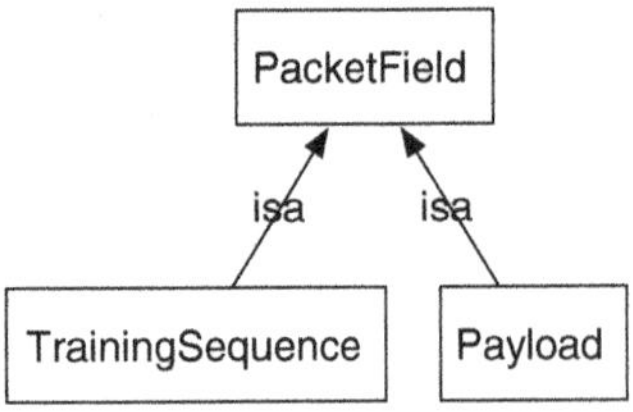

Fig. B.6 Subclasses of PacketField

Table B.2 Properties of PacketField

Property	Domain	Range
startIndex	PacketField	Integer*
filedSize	PacketField	Information

A *PacketField* class contains either user data (payload) or control information (header or trailer). The attributes of *PacketField* include *startIndex* and *fieldSize*, shown in Table B.2.

B.1.5 Signal

Signal is any time-varying or spatial varying quantity. There are different views on the classification of *Signal*. *Signal* can be divided into continuous and discrete

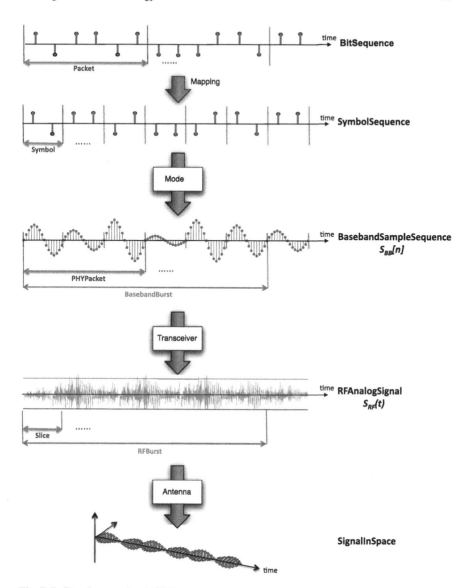

Fig. B.7 Signal processing in SDR

signal; then further divided into quantized signal and unquantized signal. However, since our ontology is developed for the cognitive radio domain, almost all the signal processing (before the DAC or the amplifier) is implemented in software. Figure B.7 shows an example of signal processing steps in a cognitive radio.

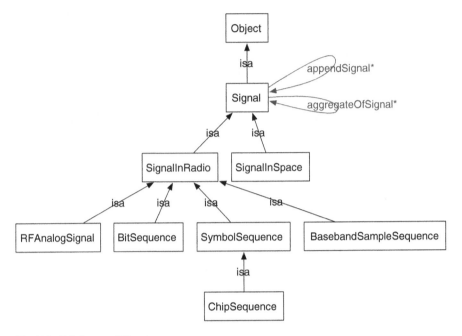

Fig. B.8 Subclasses of Signal

1. A *BitSequence* instance is generated, for instance, by a channel encoder.
2. The *BitSequence* is grouped into codewords, one for each symbol to be transmitted. The sequence of codewords is represented as *SymbolSequence*.
3. *SymbolSequence* is mapped to the amplitudes of the *I* and *Q* signals, and then multiplied by the baseband frequency to produce the *BasebandSampleSequence*.
4. The *BasebandSampleSequence* is then processed in the transceiver subsystem, producing the *RFAnalogSignal*.
5. Finally, the *RFAnalogSignal* is transmitted to the air by the antenna, becoming the *SignalInSpace*.

In our ontology, we divide the Signal class into (1) *SignalInRadio*, (2) *SignalInSpace*.

SignalInRadio and *SignalInSpace* are disjoint classes.

SignalInRadio can be further divided into (1) *BitSequence*, (2) *SymbolSequence*, (3) *BandbandSampleSequence*, and (4) *RFAnalogSignal*. Note that *Chip* is a special type of *Symbol* and thus *ChipSequence* is a subclass of *SymbolSequence*.

Signal class has the following basic properties: (1) Part of a *Signal* is also a *Signal*. (2) A *Signal* can be appended to another *Signal*, producing a new *Signal*. (3) *signalDuration*, and (4) *signalRate*.

The subclasses of Signal Class are shown in Fig. B.8. The properties of Signal and its subclasses are shown in Table B.3.

Table B.3 Properties of signal

	Properties and sub-properties		Domain	Range
Properties of Signal	signalDuration		Signal	Time
	signalRate	grossBitRate netBitRate throughput goodput symbolRate sampleRate chipRate	Signal	SignalRate
		... ...		
Properties of SignalInSpace	signalPower		SignalInSpace	Power
	signalPowerDensity		SignalInSpace	PowerDensity
	signalStrength		SignalInSpace	ElectricFieldStrength

B.1.6   Burst

Burst is a segment of *Signal*. For radio transmission, the transmit channel of the transceiver subsystem up-converts bursts of *BasebandSampleSequence* to bursts of *RFAnalogSignal*. The transmit channel consumes the coming signal burst; stores the result in a buffer; then performs the up-conversion in real time. Note that *Burst* is NOT disjoint with either*SignalInRadio* or *SignalInSpace*.

A burst of *BasebandSampleSequence* is called *BasebandBurst*, consisting of several *Packets*. The length of a *BasebandBurst* must be a multiple of the length of a physical layer packet.

A burst of *RFAnalogSignal* is called *RFBurst*, consisting of several *Slices*. Since a *Slice* corresponds to a *Symbol*, the length of a *Slice* equals to the length of the corresponding *Symbol*. The length of an *RFBurst* must be a multiple of the length of a *Slice*.

The illustrations of *BandbandBurst* and *RFBurst* are shown in Fig. B.9.

The relationships among *Burst*, *Signal* and *Packet* are shown in Fig. B.10.

The properties of *Burst* are shown in Table B.4, including: (1) *burstStartTime*, (2) *burstStopTime*, and (3) *burstLength*.

B.1.7   Sample

A *Sample* refers to a value taken at a point in time or space. A *Signal* is an aggregation of *Samples*. The properties of *Sample* include: (1) sample value, and (2) the time at which the sample is taken, shown in Fig. B.11.

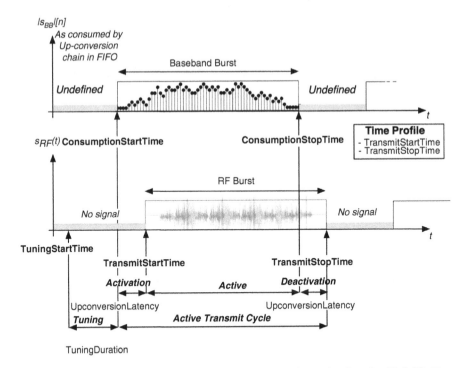

Fig. B.9 Illustration of BasebandBurst and RFBurst (*source*: Transceiver Interface Task Working Group 2009)

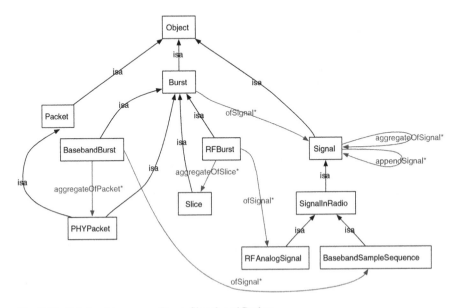

Fig. B.10 Relationships among Burst, Signal, and Packet

Table B.4 Properties of Burst

Property	Domain	Range
burstStartTime	Burst	Time
burstStopTime	Burst	Time
burstLength	Burst	Time
ofSignal	Burst	Signal

Fig. B.11 Properties of Sample

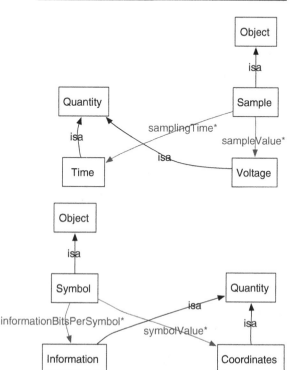

Fig. B.12 Properties of Symbol

B.1.8   Symbol

The term *Symbol* is ambiguous in the sense that it is used to mean different things. (1) *Symbol* may refer to the physically transmitted signal that is placed on the channel. It is a state of the communication channel that persists for a fixed period of time (National Communications System Technology & Standards Division 1996; Bell 1962). For example, in passband transmission a *Symbol* usually refers to a sine wave tone, whereas in baseband transmission a symbol usually refers to a pulse rather than a sine wave tone. (2) *Symbol* may be used at a higher level and refers to one information bit or a block of information bits that will be modulated using a conventional modulation scheme such as QAM (National Communications System Technology & Standards Division 1996; Bell 1962).

In this ontology, *Symbol* refers to the first definition mentioned above. The SymbolSequence described in Sect. B.1.5 also refers to the first definition.

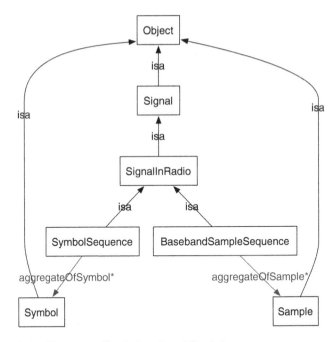

Fig. B.13 Relationships among Signal, Sample and Symbol

The properties of *Symbol* include: (1) *InformationBitsPerSymbol*, and (2) *SymbolValue*, shown in Fig. B.12.

informationBitsPerSymbol refers to the number of information bits that a symbol conveys. For example, in a differential Manchester line coding, each information bit is represented by two symbol pulses, therefore, in this case, the value of InformationBitsPerSymbol equals to $\frac{1}{2}$.

symbolValue refers to the co-ordinates of a symbol on the constellation diagram. Examples are described in Sect. B.1.1. Note that the *symbolValue* can be a complex number or a real number, depending on which modulation scheme is used. For example, in QAM modulation, the *symbolValue* is a complex number.

In summary, the relationships among *Signal*, *Symbol* and *Sample* are shown in Fig. B.13.

B.1.9   PNCode

PNCode refers to the pseudo noise code that has a spectrum similar to a random sequence of bits but is deterministically generated. *PNCode* is usually used in a direct-sequence spread spectrum system. Examples of *PNcode* include *Maximal-LengthSequences*, *GoldCode*, *KasamiCode*, *BarkerCode* and *WalshCode*, shown in Fig. B.14.

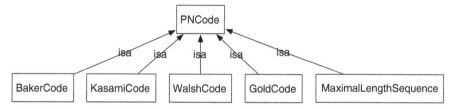

Fig. B.14 Subclasses of PNCode

B.1.10   Component

B.1.10.1   Classification of Component

A *Component* is a self-contained part of a larger entity. It often refers to a manufactured object or a software module.

A *Component* can be part of a larger component and it can have another component(s) as its subcomponent(s). Therefore, a component has two properties: (1) *isSubComponentOf*, and (2) *hasSubComponent*.

Some components can be decomposed into smaller subcomponents whereas some components can not be decomposed. In our ontology, the components that cannot be decomposed are *BasicComponent*, e.g. Multiplier and Adder.

Radio is a complex component that consists of other *RadioComponents*.

In the current version, *Component* has three subclasses: (1) *BasicComponent*, (2) *Radio*, and (3) *RadioComponent*.

Note that *RadioComponent* can NOT have *Radio* as its sub-component, this is represented as a restriction in our ontology. In addition, *FM3TRRadio* is a special type of radio that can be used as a test case of our ontology; we include *FM3TRRadio* as a sub-class of *Radio*.

The hierarchy of subclasses of the *Component* class are shown in Table B.5.

B.1.10.2   Structure of Component

The structure of a *Component* describes the structure of the function between input variables and output variables. The function is described as a set of blocks that are connected by ports. Figure B.15 shows an example of the physical layer structure component of an FM3TR radio (Willink 2000).

Each *Component* has input and output ports. One component is connected to another component by ports. *Port* is an *Object*. A *Port* can be connected to another *Port* if the two ports are carrying the same type of signal. For example, if a modulator takes a digital signal as the input and outputs an analog signal, then the output port of this modulator can be connected to another port that also carries an analog signal.

Table B.5 Subclasses of Component

Class	Subclasses			
Component	BasicComponent	Multiplier		
		Adder		
	Radio	FM3TRRadio		
	RadioComponent	ChannelDecoder		
		ChannelEncoder		
		ChannelEstimator		
		Demodulator		
		Detector	Location-Detector	
			SignalDetector	ContinuousSignal-Detector
				PulseSignalDetector
			TimeDetector	
		Equalizer		
		FM3TRComponent	Fm3tr_Dlc	
			Fm3tr_Hci	
			Fm3tr_Nwk	
			Fm3tr_Phl	
		Modulator		
		PowerAmplifier		
		Receiver		
		SourceDecoder		
		SourceEncoder		
		Transceiver		
		Transmitter		
		Receiver		
		WaveformApplication		

Figure B.16 shows an example of how to represent the structure of a component. Suppose component *C* has three sub-components (*C1*, *C2*, and *C3*), one input port (*P1*) and two output ports (*P9*, *P10*). First, the relationships between a component and its ports are modeled using the object-type property *hasPort*. The *hasPort* property has two sub-properties: *hasInputPort* and *hasOutputPort*. For instance, ¡C1 hasInputPort p2¿, ¡C1 hasOutputPort P3¿, ¡C1 hasOutputPort P4¿. Second, the relationships between *Ports* are modeled using the object-type property *isConnectedTo*. For instance, ¡P3 isConnectedTo P5¿. Note that ports can be connected only if their port types are the same. However, we have not represented this restriction in OWL.

The relationships between *Component* and *Port* are shown in Fig. B.17.

Example: OWL Representation of FM3TR Structure

Figure B.18 shows the top-level structure of an FM3TR radio (Willink 2000). Figure B.19 shows the OWL representation of this structure.

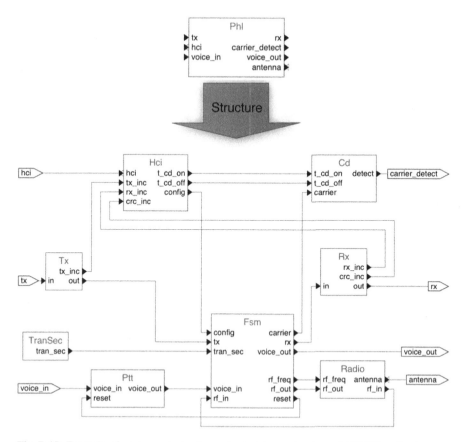

Fig. B.15 Example of component structure: physical layer structure of FM3TR radio (*source*: Willink 2000)

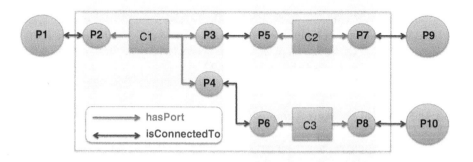

Fig. B.16 Representation of component structure

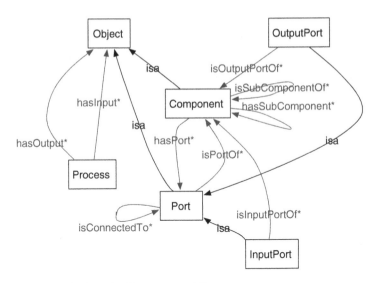

Fig. B.17 Relationships between Component and Port

B.1.10.3 Behavior of Component

The behavior of a radio component is usually described by a behavior model, e.g. PetriNet or State Transition Diagram (STD). Figure B.20 shows the structure and the behavior model of an FM3TR Physical Layer component (Willink 2000). The representation of *BehaviorModel* can be found in Sect. B.2.11. The relationship between a component and its behavior model is shown in Fig. B.21.

B.1.10.4 Capabilities of Component

A *Component* is capable of performing particular tasks, such as receiving signal or detecting spectrum opportunities. The capabilities of a *Component* are a set of processes. Therefore, we use an object-type property *hasCapability* to represent the capabilities of a *Component*, shown in Fig. B.22.

B.1.10.5 API of Software Component

A physical component contains input and output ports. A software component can implement a set of APIs to enable the interaction with other software components. Once a software component implements an API, other software components can use this API.

The relationship between component and API is shown in Fig. B.23.

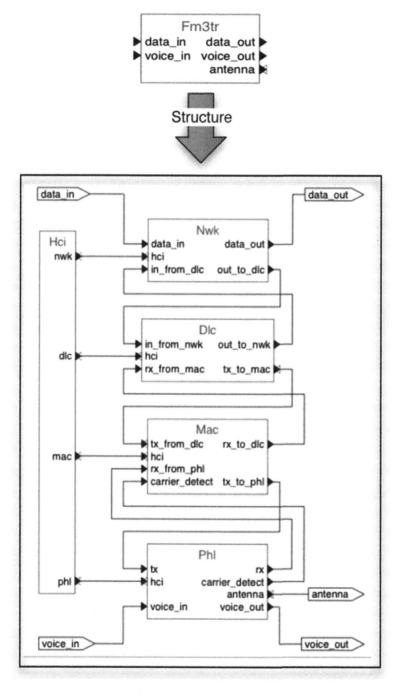

Fig. B.18 Top-level component structure of FM3TR radio (*source*: Willink 2000)

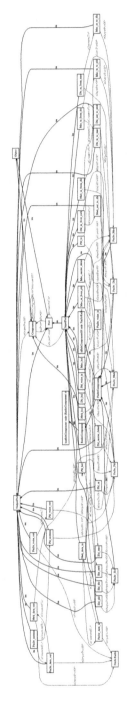

Fig. B.19 OWL representation of FM3TR radio

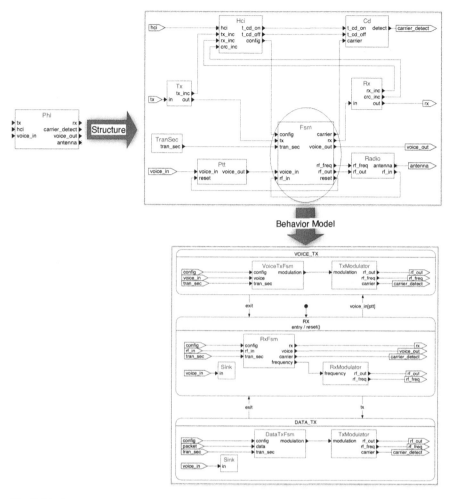

Fig. B.20 Structures and behavior model of FM3TR physical layer component (*source*: Willink 2000)

Fig. B.21 Relationships between Component and BehaviorModel

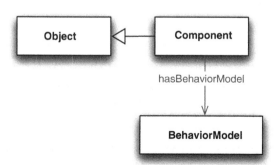

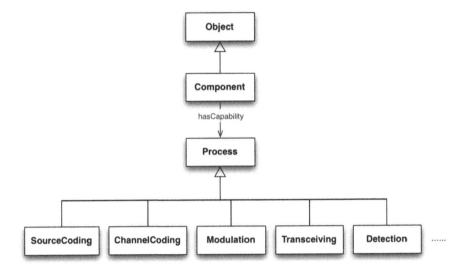

Fig. B.22 Capabilities of component

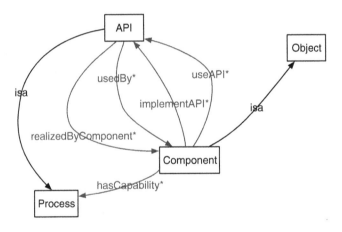

Fig. B.23 Relationships between Component and API

Example: OWL Representation of APIs of Transmitter

The API of Transmitter is specified in the "Transceiver Facility Specification" document by the Transceiver Working Group of SDR forum. Figure B.24 shows an overview of the transmitter APIs (Transceiver Interface Task Working Group 2009). Figure B.6 shows the APIs between Transmitter and Waveform Application. Tables B.6 and B.7 show the overview of each API of Transmitter (Transceiver Interface Task Working Group 2009).

Figure B.25 shows the OWL representation of the Transmitter API. The details of API and Method are shown in Sect. 2.

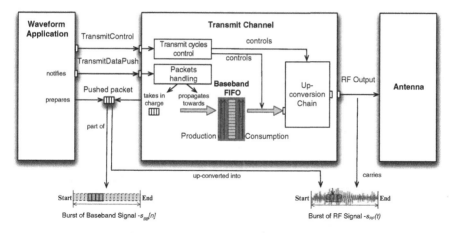

Fig. B.24 Overview of Transmitter API (*source*: Transceiver Interface Task Working Group 2009)

B.1.10.6   NetworkMembership of Component

A *Component* may have membership in a *Network*. Therefore each *Component* is associated with one or more than one *NetworkMembership*. The relationships among *Component*, *Network*, *NetworkMembership* and *Role* are shown in Sect. B.1.13.

B.1.11   *TransceiverPreset, Transfer Functions and Constraints of Transfer Functions*

B.1.11.1   TransceiverPreset

TransceiverPreset refers to a set of tunable parameters that are provided with corresponding requested values before the up-/down-conversion is activated. According to Transceiver Interface Task Working Group (2009), *TransceiverPreset* is composed of the tunable parameters of *BasebandSampleSequence*, *ChannelMask*, *GroupDelayMask* and *SpectrumMask*.

B.1.11.2   TxChannelTransferFunction and RxChannelTransferFunction

TxChannelTransferFunction refers to the transfer function response of the transformation operated by up-conversion chain between the *BasebandSampleSequence* and *RFAnalogSignal*.

RxChannelTransferFunction refers to the transfer function response of the transformation operated by the down-conversion chain between the *RFAnalogSignal* and *BasebandSampleSequence*.

Table B.6 Transmitter API (1): TransmitControl (*source*: Transceiver Interface Task Working Group 2009)

Signature summary (pseudo-code)	Used by	Realized by	Description

createTransmitCycleProfile(
    Time requestedTransmitStartTime,
    Time requestedTransmitStopTime,
    UShort requestedPresetId,
    Frequency requestedCarrierFrequency,
    AnaloguePower requestedNominalRFPower)
	Waveform Application	Transceiver Subsystem	Creation of a Transmit Cycle Profile

ConfigureTransmitCycle(
    Ulong targetCycleId,
    Time requestedTransmitStartTime,
    Time requestedTransmitStopTime,
    Frequency requestedCarrierFrequency,
    AnaloguePower requestedNominalRFPower)
	Waveform Application	Transceiver Subsystem	Configuration of an existing Transmit Cycle Profile

setTransmitStopTime(
    Ulong targetCycleId,
    Time requestedTransmitStopTime)
	Waveform Application	Transceiver Subsystem	Specification of the end time of a Transmit Cycle

Table B.7 Transmit API (2): TransmitDataPush (*source:* Transceiver Interface Task Working Group 2009)

Signature summary (pseudo-code)	Used by	Realized by	Description
pushBBSamplesTx(BBPacket thePushedPacket Boolean endOfBurst)	Waveform Application	Transceiver Subsystem	Notifies availability of a baseband samples packets

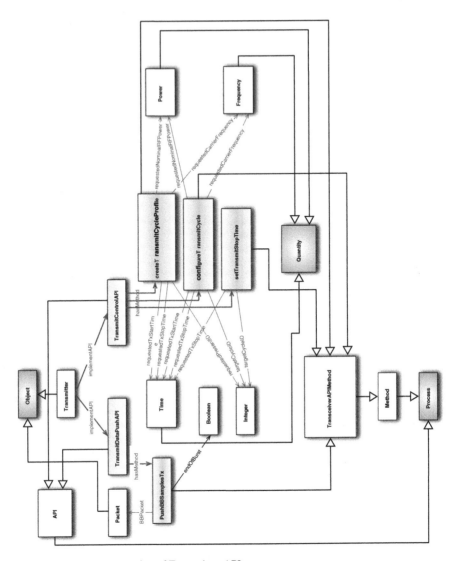

Fig. B.25 OWL representation of Transmitter API

TxChannelTransferFunction is used to characterize the process of *Transmitting* and *RxChannelTransferFunction* is used to characterize the process of *Receiving*.

B.1.11.3 ChannelMask, SpectrumMask, and GroupDelayMask

ChannelMask, SpectrumMask and *GroupDelayMask* are the constraints of the *Tx-ChannelTransferFunction* and *RxChannelTransferFunction* (Transceiver Interface Task Working Group 2009).

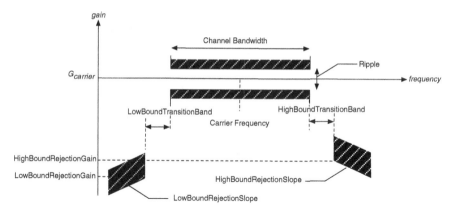

Fig. B.26 Characteristics of SpectrumMask (*source*: Transceiver Interface Task Working Group 2009)

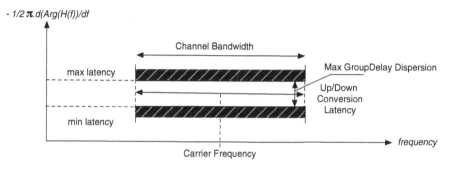

Fig. B.27 Characteristics of GroupDelayMask (*source*: Transceiver Interface Task Working Group 2009)

ChannelMask refers to the requirements that shall be met by the *ChannelTransferFunction* of a given conversion chain.

SpectrumMask is used to characterize the spectrum mask to be satisfied by the modulus of the *ChannelTransferFunction*. Figure B.26 shows the characteristics of *SpectrumMask*.

GroupDelayMask is used to characterize the group delay response to be satisfied by the *ChannelTransferFunction*. Figure B.27 shows the characteristics of *GroupDelayMask*.

Table B.8 shows the overview of the properties associated to *ChannelMask*, *SpectrumMask* and *GroupDelayMask* (Transceiver Interface Task Working Group 2009).

B.1.11.4   Summary of Transmitter-Related Classes

Figure B.28 shows the relationships among the classes related to *Transmitter*. *TransmitterPreset* and *Transmitter* are objects; they both participate in the process of

Table B.8 Properties of ChannelMask, SpectrumMask and GroupDelayMask

	Properties	Domain	Range
ChannelMask	carrierFrequencyAccuracy	ChannelMask	Frequency
	channelBandwidth	ChannelMask	Frequency
SpectrumMask	highBoundRejectionGain	SpectrumMask	Decibel
	highBoundRejectionSlope	SpectrumMask	GainSlope
	highBoundTransitionBand	SpectrumMask	Frequency
	lowBoundRejectionGain	SpectrumMask	Decibel
	lowBoundRejectionSlope	SpectrumMask	GainSlope
	lowBoundTransitionBand	SpectrumMask	Frequency
	ripple	SpectrumMask	Decibel
GroupeDelayMask	maxGroupDelayDispersion	GroupDelayMask	Frequency

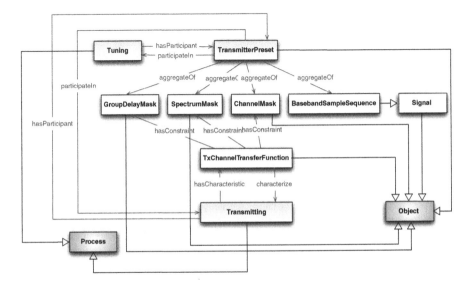

Fig. B.28 Characteristics of Transceiver

Tuning and Transmitting. The details of *Tuning* and *Transmitting* will be described in Sects. B.2.3 and B.2.4.

B.1.11.5 Properties of Transmitter

The properties of *Transmitter* class is summarized in Table B.9 (Transceiver Interface Task Working Group 2009).

B.1.12 Detector and DetectionEvidence

A *Detector* is a device that can detect three types of DetectionEvidence: (1) TimeEvidence, (2) LocationEvidence, and (3) SignalEvidence.

Table B.9 Properties of Transmitter

Property	Domain	Range
basebandFIFOSize	Transmitter	Integer*
basebandCodingBits	Transmitter	Integer*
basebandNominalPower	Transmitter	Power
carrierFrequency	Transmitter	Frequency
consumptionStartTime	Transmitter	Time
consumptionStopTime	Transmitter	Time
maxCycleId	Transmitter	Integer*
maxOnTime	Transmitter	Time
maxPushedPacketSize	Transmitter	Integer*
maxTransmitDataPushInvocationDuration	Transmitter	Time
maxTuningDuration	Transmitter	Time
maxTxCycleProfiles	Transmitter	Integer*
maxUpconversionLatency	Transmitter	Time
minOffTime	Transmitter	Time
minPacketStorageAnticipation	Transmitter	Time
minReactivationTime	Transmitter	Time
minTransmitStartAnticipation	Transmitter	Time
minTransmitStartProximity	Transmitter	Time
nominalRFPower	Transmitter	Power
overflowMitigation	Transmitter	String*
reactivationTime	Transmitter	Time
transmissionPower	Transmitter	Power
transmitCycle	Transmitter	Integer*
transmitStartTime	Transmitter	Time
transmitStopTime	Transmitter	Time
transmitTimeProfileAccuracy	Transmitter	Time
tuningDuration	Transmitter	Time
tuningStartThreshold	Transmitter	Integer*
tuningStartTime	Transmitter	Time
upconversionLatency	Transmitter	Time

The properties of Detector and its subclasses are shown in Table B.10.

The relationship between Detector and DetectionEvidence is shown in Fig. B.29. The properties of DetectionEvidence and its subclasses are shown in Table B.11.

B.1.13 Network, Network Membership and Role

A *Component* may act as a member in a *Network*. Each *NetworkMembership* is (1) associated with one *Component*, (2) belongs to a *Network*, and (3) has its *Role* in the Network. The *Role* of a member can be *master*, *slave* or *peer*. The relationships among *Network*, *NetworkMembership*, *Role* and *Component* are shown in Fig. B.30.

Table B.10 Properties of Detector and its subclasses

	Properties	Domain	Range
Detector	scanDuration	Detector	Time
	scanInterval	Detector	Time
	detectEvidence	Detector	DetectionEvidence
LocationDetector	detectEvidence	LocationDetector	LocationEvidence
SignalDetector	detectEvidence	SignalDetector	SignalEvidence
	endFrequency	SignalDetector	Frequency
	rssi	SignalDetector	Power
	sampleRate	SignalDetector	SampleRate
	setToDetect	SignalDetector	Signal
	signalDetectionPrecision	SignalDetector	Voltage
	signalDetectionThreshold	SignalDetector	Voltage
	signalToNoiseRatio	SignalDetector	Decibel
	startFrequency	SignalDetector	Frequency
TimeDetector	detectEvidence	TimeDetector	TimeEvidence

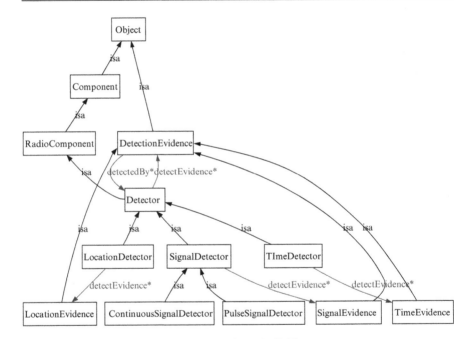

Fig. B.29 Relationships between Detector and DetectionEvidence

B.1.14   Agent and Goal

Agent is a special type of *Object*. The definition of *Agent* varies in different domains. In artificial intelligence, *Agent* refers to an autonomous entity which observes and acts upon an environment and directs its activity towards achieving its own goals

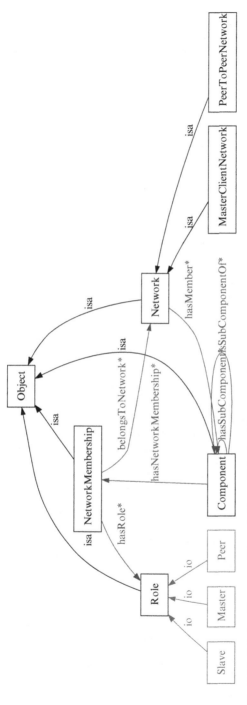

Fig. B.30 Relationships among Network, NetworkMembership, Role and Component

Table B.11 Properties of DetectionEvidence and its subclasses

	Properties	Domain	Range
Property of DetectionEvidence	confidence	DetectionEvidence	Percentage
	timeStamp	DetectionEvidence	Time
Property of LocationEvidence	location	LocationEvidence	Location
Property of TimeEvidence	time	TimeEvidence	Time
Property of SignalEvidence	consecutiveEmptyScanCount	SignalEvidence	Integer*
	detectedSignal	SignalEvidence	Signal
	lastCompleteEmptyScanDuration	SignalEvidence	Signal
	lastCompleteEmptyScanTime	SignalEvidence	Time
	lastDetectionTime	SignalEvidence	Time
	peakSensedPower	SignalEvidence	Power
	sensedEndFrequency	SignalEvidence	Frequency
	sensedStartFrequency	SignalEvidence	Frequency

(Russell and Norvig 2003). The essences of an agent includes: (1) sensing of and reaction to the environment, i.e. an agent is able to sense the environment and react properly to the changes of the environment; (2) autonomy, i.e. an agent can perform a task without human intervention; (3) persistency, i.e., for example, if a software program is an agent, then it should be executed continuously over time rather than invoked on demand and terminates after the completion of its function; (4) goal-directed, an agent should be capable of choosing among multiple options and select the one that can achieve the goal (Franklin and Graesser 1997).

The above properties distinguish an agent from an ordinary software program or module. In the domain of cognitive radio, a radio component has inputs and outputs. It performs tasks on its own by running a predefined algorithm. It could be said that the radio component senses the environment via the inputs and responds to the environment via outputs. In this sense, the radio component is capable of reacting to the environment and has some degree of autonomy. However, a radio component may be invoked for once and then goes into an idle state, waiting to be invoked again. In this sense, this radio component does not satisfy the temporal persistency property. Furthermore, in order to become an agent, a radio component must have goal-directed behavior, i.e. it does not simply sense and react upon the environment autonomously (Franklin and Graesser 1997), it must be able to achieve a set of goals, e.g. avoid detection and interference, maximize throughput, etc.

DOLCE has a clear classification of *Agent*, i.e. which object is agentive and which is non-agentive. In this ontology, we do not restrict any of the radio components as a subclass of *Agent*. Instead, we define that an *Object* is an *Agent* if and only if it has a *Goal*. Given such a necessary and sufficient condition, it can be inferred whether a radio component is or is not an agent. The subclasses of the *Goal* class and the relationships between *Agent* and *Goal* are shown in Fig. B.31.

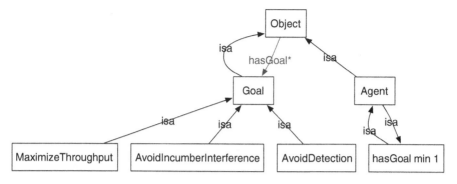

Fig. B.31 Relationships between Agent and Goal

B.2 Process

B.2.1 AIS and Protocol

Air Interface Specification (AIS) is closely related to the term Waveform. The P1900.1 working group defines "waveform" as follows (DYSPAN P1900.1 Working Group 2008):

(a) The set of transformations and protocols applied to information that is transmitted over a channel and the corresponding set of transformations and protocols that convert received signals back to their information content.
(b) The time-domain or frequency-domain representation of an RF signal.
(c) The representation of transmitted RF signal plus optional additional radio functions up to and including all network layers.

AIS is the specification of a set of processes that are applied to the transmitted and received information. For instance, if two radios want to communicate with each other, the signals provided by the two radios must both satisfy the AIS, whereas the details of implementation may be different. In this sense, AIS is equivalent to the term Waveform defined in (a).

As it is discussed in Sect. 8.2.2.2, the specification of AIS is an *Object* whereas the implementation of AIS is a *Process*. In the current version, AIS refers to the implementation, thus it is a *Process*.

Typically, AIS is layered, with interfaces defined for each layer. Each layer consists of one or more protocols that perform the layer's functionality. A protocol defines the format and the ordering of messages exchanged between two or more communicating entities, as well as the actions taken on the transmission and/or receipt of a message or other event.

For example, in cdma2000 1xEV-DO (3rd Generation Partnership Project 2 (3GPP2) 2000), the AIS is divided into several layers, shown in Fig. B.32. The protocols defined for each layer are shown in Fig. B.33. The MAC layer consists of

Fig. B.32 Air interface
layering architecture (*source*:
3rd Generation Partnership
Project 2 (3GPP2) 2000)

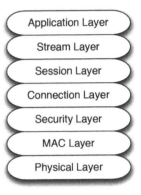

Application Layer

Stream Layer

Session Layer

Connection Layer

Security Layer

MAC Layer

Physical Layer

multiple protocols such as *Control Channel Protocol* and *Forward Traffic Channel Protocol*. Hence, AIS is an aggregation of protocols. From another point of view, AIS is also an aggregation of various processes, i.e. AIS provides the specification for modulation, channel coding, source coding, etc. In our ontology, we only focus on the physical layer, data link layer and network layer of the AIS. The relationships among AIS, Protocol and Process are illustrated in Fig. B.34.

B.2.2 API and Method

A general discussion of *API* and *Method* was already shown in Sects. 8.2.3.3 and B.1.10.5.

B.2.3 Tuning

Tuning refers to the process of setting the parameters of a radio to requested values. The relationships between *Tuning* and other classes related to *Transmitter* are shown in Sect. B.1.11.4.

B.2.4 Transmitting

Transmitting refers to the process of up-converting bursts of *BasebandSampleSequence* to bursts of *RFAnalogSignal*. The *Transmitter* consumes the coming signal burst; stores the result in a buffer; then performs the up-conversion in real time.

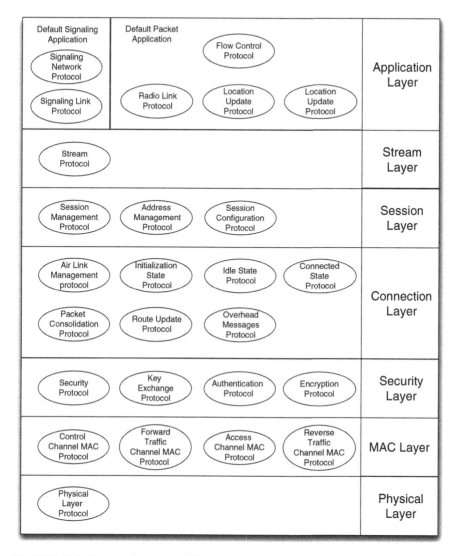

Fig. B.33 Default protocols of cdma2000 1xEV-DO (*source*: 3rd Generation Partnership Project 2 (3GPP2) 2000)

B.2.5   Receiving

Receiving refers to the process of down-converting bursts of *RFAnalogSignal* to bursts of *BasebandSampleSequence*.

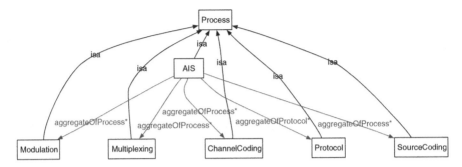

Fig. B.34 Relationships among AIS, Protocol and Process

B.2.6   SourceCoding

SourceCoding refers to the process of encoding information using fewer bits. Source coding helps reducing the consumption of hard disk or transmission bandwidth.

B.2.7   ChannelCoding

ChannelCoding refers to the process of adding redundancy to the data at the transmitter in order to correct data error and thus lower the error rate at the receiver.

B.2.8   Modulation

The relationships of *Modulation*, *Modulator* and *Alphabet* were already discussed in Sect. 8.2.2.2.

   In general, *Modulation* is a process that takes a digital signal as input and converts it to an analog signal. Then the analog signal is transmitted to the wireless channel. The changes in the carrier signal are chosen from a finite number of M alternative symbols, which is called *alphabet*.

B.2.9   Multiplexing

Multiplexing refers to the process of combining multiple signals or information streams into one signal and sending it through a shared medium.

Fig. B.35 Subclasses of
BehaviorModel

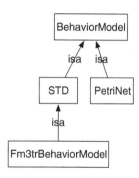

B.2.10 PNSequenceGeneration

PNSequenceGeneration refers to the process to generate the *PNCode*. *PNCode* can
be used in scrambler and spectrum spreading.

B.2.11 BehaviorModel

The behavior of a radio component is usually described by a behavior model, e.g.
PetriNet or *State Transition Diagram* (STD). The FM3TR specification describes
the behavior using *State Transition Diagrams* (STD). Thus, in our ontology,
FM3TRBehaviorModel is a subclass of STD. The subclasses of *BehaviorModel* are
shown in Fig. B.35.

The basic elements of STD include (1) *State*, (2) *Transition* between states, (3)
Action that is triggered by the state transition, (4) *Activity* that consists of a sequence
of actions, (5) *Event* that triggers a state transition.

B.2.11.1 State

"A state models a situation during which some invariant condition holds" (Object
Mangement Group 2007). It has three properties: (1) *doAction*, (2) *isFinal*, and (3)
isInitial.

isFinal and *isInitial* are Boolean data type properties. The initial state is the state
that an object is in when it is first created, whereas a final state is one in which no
transitions lead out of. Also, a state transition will trigger an action, or a sequence
of actions, thus a state is associated with an action by property *doAction*.

Fig. B.36 State transition table

CurrentState (right)	StateA	StateB	StateC
Event that causes the transition (below)			
EventX	...	...	...
EventY	...	StateC	...
EventZ	...	...	...

B.2.11.2 Transition

"A transition is a directed relationship between a source vertex and a target vertex. It may be part of a compound transition, which takes the state machine from one state configuration to another, representing the complete response of the state machine to an occurrence of an event of a particular type" (Object Mangement Group 2007). In other words, a transition is a change from one state to another. Thus, a *Transition* is associated with a target state and a source state. As mentioned above, a state transition will trigger an action or a sequence of actions. Therefore, a *Transition* is associated with an *Action* by property *cause*. On the other hand, a transition is usually triggered by an event that is either internal or external to the object. Hence, a *Transition* is associated with *Event* by property *casuedBy*. Note that in our ontology, the distinctions among *Event*, *Condition* and *Guard* are not expressed. In summary, the Transition class has four properties: (1) *cause*, (2) *casuedBy*, (3) *sourceState* and (4) *targetState*.

B.2.11.3 Action, Activity and Event

The distinction between *Action* and *Activity* is described in the UML superstructure specification (Object Mangement Group 2007). In short, an action is performed when a state transits to another state. An activity consists of a sequence of actions. "Each action in an activity may execute zero, one, or more times for each activity execution" (Object Mangement Group 2007). In our ontology, *Activity* is equivalent to *Process* and thus is not represented as a separate class. The relationship between action and activity is modeled in the following way. First, *Action* is modeled as a subclass of *Process*. Second, a *Process* is an aggregation of *Action*. Third, an *Action* can be appended to another *Action*. A sequence of actions forms an *Activity* (Process).

In UML, an *Event* is a notable occurrence at a particular point in time. A state transition is triggered by an internal or external event.

B.2.11.4 State Transition Diagram (STD)

In this ontology, we only focus on the finite state machine (FSM) to represent behaviors. An FSM can be described using a state transition table, as shown in

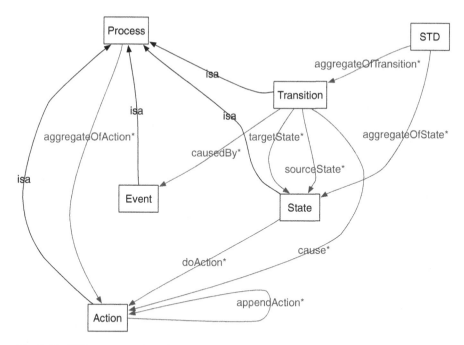

Fig. B.37 OWL representation of state transition diagram

Fig. B.36. It can be seen that a *State Transition Diagram* can be viewed as an aggregation of *State* and *Transition*. The OWL representation of STD and the relationships among STD, State, Transition, Event, Action, and Activity (Process) are shown in Fig. B.37.

B.2.11.5   Example: OWL Representation of Physical Layer FSM FM3TR Radio

In this section, an example from the FM3TR specification is used to illustrate how to use the approach described above to represent a Finite State Machine (FSM). Figure B.38 shows a physical layer FSM specification of a FM3TR radio. This radio operates in a half-duplex mode. At the beginning, the radio idles at the RX state. When the PTT (push to talk) button is pressed and voice comes in from the Voice_Tx port, the radio transits from the RX state to the VOICE_Tx state. When it is finished, the radio will transit back to the RX state and reset the PTT. On the other hand, when there is data coming in from the TX port, the radio will transit from the Rx state to the DATA_TX state; when it is finished, the radio will transit back to the RX state and again reset the PTT (Transceiver Interface Task Working Group 2009). The diagram shows that each state specifies distinct receive and transmit activity. The realization of this FSM is shown in Fig. B.39.

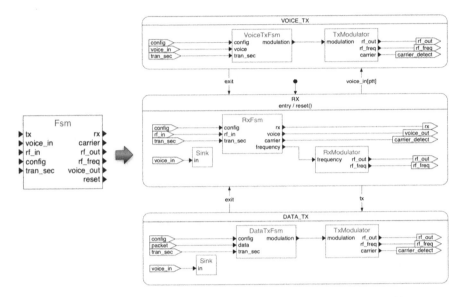

Fig. B.38 Physical Layer FSM Specification of a FM3TR Radio (*source:* Willink 2000)

B.3   Value

In this ontology, *value* refers to the magnitude of the property of either an object or a process. It is represented as a top-level class. The subclasses of *Value* class include (1) CartesianCoordinates, (2) ComplexValue, (3) FloatValue, and (4) IntegerValue.

The properties of each subclass are shown in Table B.12.

B.4   Quantity and UnitOfMeasure

As we discussed in Sect. 8.2.1 in Chap. 8, *Quantity* plays a similar role as *Quality* in DOLCE, which are the basic attributes or properties that can be perceived or measured. Quantity carries three types of information: the type of the quantity (e.g., mass, length), the magnitude of the property (typically a real or integer number) and the unit of measurement associated with the given magnitude (e.g., [*kg*], [*m*]). Quantity is a top-level class in this ontology and can be further divided (sub-classified) into different types, such as length, frequency, time, etc. Each quantity is associated with a unit and a value. *UnitOfMeasure* is also a top-level class in this ontology; it represents the unit of a quantity. The summary of *Quantity* and *UnitOfMeasure* is shown in Table B.13.

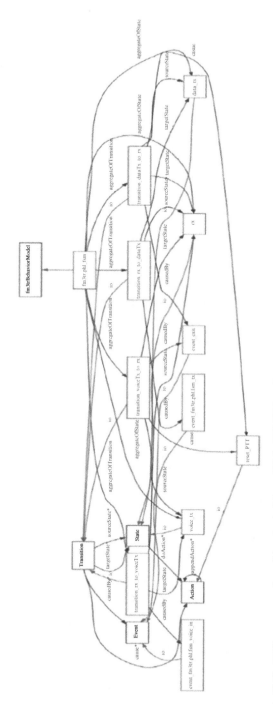

Fig. B.39 OWL representation of the physical layer FSM of FM3TR radio

Table B.12 Properties of Value class and its subclasses

Properties	Domain	Range
hasPrecision	Value	Integer*
hasX	CartesianCoordinates	Float*
hasY	CartesianCoordinates	Float*
hasZ	CartesianCoordinates	Float*
hasImg	ComplexValue	Float*
hasReal	ComplexValue	Float*
hasFloat	FloatValue	Float*
hasInt	IntegerValue	Integer*

Table B.13 Overview of Quantity and UnitOfMeasure

Quantity			UnitOfMeasure
Bandwidth	ComputingBandwidth		bit/s
	SignalProcessBandwidth		Hz
Coordinates			N/A
ElectricCurrent			I
ElectricFieldStrength			dB uV/m
Energy			J
Frequency			Hz
Information			Bit
Length			m
Location	AreaLocation	LocationPolygone	N/A
	LocationPoint	Latitude	N/A
		Longitude	N/A
		Altitude	N/A
Power			dB m
PowerDensity			dB m/m^2, mW/cm^2
Ratio	Decibel		N/A
	Percentage		N/A
SignalRate	ChipRate		chip/s
	Goodput		bit/s
	GrossBitRate		bit/s
	NetBitRate		bit/s
	SampleRate		sample/s
	SymbolRate		symbol/s
	Throughput		bit/s
Time			s
Voltage			V

References

3rd Generation Partnership Project 2 (3GPP2) (2000) cdma2000 high rate packet data air interface specification, 3GPP2 C.S0024-A version 2.0. URL http://www.3gpp2.org/public_html/specs/C.S0024_v2.0.pdf

Albus JS, Pape CL, Robinson IN, Chiueh T-C, McAulay AD, Pao Y-H (1992) RCS: a reference model for intelligent control. IEEE Comput 25:56–79

Amanna A, Reed J (2010) Survey of cognitive radio architectures. In: Proceedings of the IEEE SoutheastCon 2010. IEEE, pp 292–297

Ambler SW (2004) The object primer: agile model driven development with UML2. Cambridge University Press, New York

Anderson JR, Bothell D, Byrne MD, Douglass S, Lebiere C, Qin Y (2004) An integrated theory of the mind. Psychol Rev 111:1036–1060

Antsaklis PJ, Passino KM (1992) Introduction to intelligent control systems with high degrees of autonomy. In: Passino KM, Antsaklis PJ (eds) Introduction to intelligent and autonomous control. Kluwer Academic, Norwell

Åström KJ (1989) Adaptive control. Addison-Wesley, Reading

Austin JL (1962) How to do things with words. Oxford University Press, London

Beaulieu A (2009) Learning SQL, 2nd edn. O'Reilly, Sebastopol

Bell DA (1962) Information theory and its engineering applications, 3rd edn. Pitman, New York

Blossom E (2004) GNU radio: tools for exploring the radio frequency spectrum. Linux J 2004:4. URL http://dl.acm.org/citation.cfm?id=993247.993251

Boyd J (1987) A discourse on winning and losing. Technical report, Maxwell AFB

Bray T, Paoli J, Maler E, Yergeau F, Sperberg-McQueen CM (2008) Extensible markup language (XML) 1.0 (fifth edition). W3C recommendation, W3C, retrieved September 24, 2012, from http://www.w3.org/TR/2008/REC-xml-20081126/

Clark DD, Partrige C, Ramming JC, Wroclawski JT (2003) A knowledge plane for the internet. In: Proceedings of the SIGCOMM 2003, pp 25–29

Cognitive Radio Working Group (2007) SDRF cognitive radio definitions (SDRF-06-R-0011-V1.0.0). Technical report, The Wireless Innovation Forum, Retrieved on September 24 2012, from URL http://groups.winnforum.org/d/do/1585, retrieved on September 24, 2012, from http://groups.winnforum.org/d/do/1585

Cognitive Radio Working Group (2008) Cognitive radio definition and nomenclature (SDRF-06-P-0009-V1.0.0). Technical report, The Wireless Innovation Forum, retrieved on September 24, 2012, from http://www.sdrforum.org/pages/documentLibrary/documents/SDRF-06-P-0009-V1_0_0_CRWG_Defs.pdf

Communications Research Centre Canada (2011) SCARI: software communication architecture reference implementation. Retrieved September 24, 2012, from https://www.crc.gc.ca/en/html/crc/home/research/satcom/sdr/products/scari_suite/scari_suite

S. Li and M.M. Kokar, *Flexible Adaptation in Cognitive Radios*,
Analog Circuits and Signal Processing, DOI 10.1007/978-1-4614-0968-7,
© Springer Science+Business Media, LLC 2013

Denker G, Elenius D, Wilkins D (2009) Cognitive radio policy language and policy engine. In: Fette BA (ed) Cognitive radio technology. Elsevier, Burlington (Chap 17)

Dolwin C, Zhong S (2007) Functional description language. Technical report, E2R II White Paper

Durfee EH, Lesser VR, Corkill DD (1990) Cooperation through communication in a distributed problem-solving network. In: Cognition, computing, and cooperation. Greenwood Publishing Group, Westport (Chap 7)

DYSPAN P19001 Working Group (2008) IEEE standard definitions and concepts for dynamic spectrum access: terminology relating to emerging wireless networks, system functionality, and spectrum management. Technical report, IEEE, Piscataway

Endsley M, Garland D (2000) Situation awareness, analysis and measurement. Lawrence Erlbaum Associates, Mahway

Ferdinand M, Zirpins C, Trastour D (2003) Lifting XML schema to OWL. Intell Syst Spec Issue Agents Mark 18(6):32–38

Fette BA (2009) History and background of cognitive radio technology. In: Fette BA (ed) Cognitive radio technology. Elsevier, Burlington (Chap 1)

Fette BA, Kokar MM, Cummings M (2008) Next-generation design issues in communications. Portable Des Mag 3:20–24

Finin T, Weber J, Wiederhold G, Genesereth M, Fritzson R, Mckay D, McGuire J, Pelavin R, Shapiro S, Beck C (1993) Draft specification of the KQML agent-communication language. Technical report, The DARPA Knowledge Sharing Initiative External Interfaces Working Group, Baltimore

FIPA (2002a) FIPA ACL message structure. Technical report, FIPA, retrieved on September 24, 2012, from http://www.fipa.org/specs/fipa00061/SC00061G.pdf

FIPA (2002b) FIPA brokering protocol specification. Technical report, FIPA, retrieved on September 24, 2012, from http://www.fipa.org/specs/fipa00033/SC00033H.pdf

FIPA (2002c) FIPA communicative act library specification. Technical report, FIPA, retrieved on September 24, 2012, from http://www.fipa.org/specs/fipa00037/SC00037J.pdf

FIPA (2002d) FIPA contract net protocol specification. Technical report, FIPA, retrieved on September 24, 2012, from http://www.fipa.org/specs/fipa00029/SC00029H.pdf

FIPA (2002e) FIPA iterated contract net interaction protocol specification. Technical report, FIPA, retrieved on September 24, 2012, from http://www.fipa.org/specs/fipa00030/SC00030H.pdf

FIPA (2002f) FIPA propose interaction protocol specification. Technical report, FIPA, retrieved on September 24, 2012, from http://www.fipa.org/specs/fipa00036/SC00036H.pdf

FIPA (2002g) FIPA query interaction protocol specification. Technical report, FIPA, URL http://www.fipa.org/specs/fipa00027/SC00027H.pdf, retrieved on September 24, 2012, from http://www.fipa.org/specs/fipa00027/SC00027H.pdf

FIPA (2002h) FIPA recruiting protocol specification. Technical report, FIPA, URL http://www.fipa.org/specs/fipa00034/SC00034H.pdf, retrieved on September 24, 2012, http://www.fipa.org/specs/fipa00034/SC00034H.pdf

FIPA (2002i) FIPA request interaction protocol specification. Technical report, FIPA, URL http://www.fipa.org/specs/fipa00026/SC00026H.pdf, retrieved on September 24, 2012, http://www.fipa.org/specs/fipa00026/SC00026H.pdf

FIPA (2002j) FIPA request when protocol specification. Technical report, FIPA, URL http://www.fipa.org/specs/fipa00028/SC00028H.pdf, retrieved on September 24, 2012, from http://www.fipa.org/specs/fipa00028/SC00028H.pdf

FIPA (2002k) FIPA SL content language specification. Technical report, FIPA, retrieved on September 24, 2012, from http://www.fipa.org/specs/fipa00008/

FIPA (2002l) FIPA subscribe interaction protocol specification. Technical report, FIPA, URL http://www.fipa.org/specs/fipa00035/SC00035H.pdf, retrieved on September 24, 2012, from http://www.fipa.org/specs/fipa00035/SC00035H.pdf

Franklin S, Graesser A (1997) Is it an agent, or just a program? Intell Agents 3:21–36

Genesereth MR, Fikes RE (1992) Knowledge interchange format version 3.0 reference manual. Technical report, Computer Science Department at Stanford University, retrieved on September 24, 2012, from http://www-ksl.stanford.edu/knowledge-sharing/papers/kif.ps

Giacomoni J, Sicker DC (2005) Difficulties in providing certification and assurance for software defined radios. In: First IEEE international symposium on new frontiers in dynamic spectrum access networks, DySPAN. IEEE, Baltimore, pp 526–538

Gil Y, Ratnakar V (2002) A comparison of (semantic) markup languages. In: 15th international FLAIRS conference. The Florida Artificial Intelligence Research Society

Goldberg DE (1989) Genetic algorithms in search, optimization, and machine learning, 1st edn. Addison-Wesley, Reading

Group IPW (2011) IEEE standard for policy language requirements and system architectures for dynamic spectrum access systems. IEEE Std 1900.5^{TM}-2011

Gudgin M, Hadley M, Mendelsohn N, Lafon Y, Moreau JJ, Karmarkar A, Nielsen HF (2007) SOAP version 1.2 part 1: Messaging framework, 2nd edn. W3C recommendation, W3C, retrieved September 24, 2012, from http://www.w3.org/TR/2007/REC-soap12-part1-20070427/

Haykin S (2005) Cognitive radio: brain-empowered wireless communications. IEEE J Sel Areas Commun 23:201

He A, Gaeddert J, Bae KK, Newman TR, Reed JH, Morales L, Park CH (2009) Development of a case-based reasoning cognitive engine for ieee 802.22 wran applications. SIGMOBILE Mob Comput Commun Rev 13:37–48. DOI http://doi.acm.org/10.1145/1621076.1621081, URL http://doi.acm.org/10.1145/1621076.1621081

Holland JH (1992) Adaptation in natural and artificial systems. MIT Press, Cambridge

IEEE P19005 Working Group (2011) Policy language requirements and system architecture for dynamic spectrum access systems. Technical report, IEEE P1900.5 Working Group, Piscataway

Jennings N (1996) Coordination techniques for distributed artificial intelligence. In: O'Hare G (ed) Foundations of distributed artificial intelligence. Wiley, New York (Chap 6)

Kokar MM, Lechowicz L (2009) Language issues for cognitive radio. Proc IEEE 97(4):689–707

Kokar MM, Baclawski K, Eracar YA (1999) Control theory-based foundations of self-controlling software. IEEE Intell Syst 14(3):37–45

Kokar MM, Hillman D, Li S, Fette B, Marshall P, Cummings M, Martin T, Strassner J (2008) Towards a unified policy language for future communication networks: a process. In: 3rd IEEE symposium on new frontiers in dynamic spectrum access networks, DySPAN. IEEE, Chicago

Kokar MM, Brady D, Baclawski K (2009) The role of ontologies in cognitive radios. In: Fette BA (ed) Cognitive radio technology. Elsevier, Burlington, (Chap 13)

Kolodzy P (2009) Communications policy and spectrum management. In: Fette BA (ed) Cognitive radio technology. Elsevier, Burlington (Chap 2)

Laird JE, Newell A, Rosenbloom PS (1987) Soar: an architecture for general intelligence. J Artif Intell 33:1–64

Langley P, Laird E John, Rogers S (2009) Cognitive architectures: research issues and challenges. Cogn Syst Res 10(2):141–160

Le B, Rodriguez FAG, Chen Q, Li BP, Ge F, ElNainay M, Rondeau TW, Bostian CW (2007) A public safety cognitive radio node. In: Proceedings of the 2007 SDR forum technical conference, SDRF

Li S, Kokar MM (2010) Description of cognitive radio ontology (WINNF-10-S-0007). Technical report, The Wireless Innovation Forum, Retrieved on September 24 2012, from URL http://groups.winnforum.org/d/do/3370, retrieved on September 24 2012, from http://groups.winnforum.org/d/do/3370

Li S, Kokar MM, Moskal J (2008) Policy-driven ontology-based radio: A public safety use case. In: Software defined radio technical conference SDR'08. The Software Defined Radio Forum, Washington

Li S, Kokar MM, Moskal J (2011) An implementation of collaborative adaptation of cognitive radio parameters using an ontology and policy based approach. Analog Integr Circuits Signal Process 69(2-3):283–296

Mahmud QH (2007) Cognitive networks: towards self-aware networks. In: Mahmud QH (ed) Cognitive Networks: towards self-aware networks. Wiley, West Sussex

Mannion P (2006) Smart radios stretch spectrum. Technical report, Electronic Engineering Times (EETimes), retrieved on September 24, 2012, from http://www.eetimes.com/electronics-news/4057092/Smart-radios-stretch-spectrum

Markosian N (2000) What are physical objects. Philos Phenomenol Res 61:375–395

Marshall P (2009) Spectrum awareness and access considerations. In: Fette BA (ed) Cognitive radio technology. Elsevier, Burlington (chap 5)

Masolo C, Borgo S, Gangemi A (2009) Dolce: a descriptive ontology for linguistic and cognitive engineering. Technical report, Institute of Cognitive Science and Technology, Italian National Research Council

Matheus CJ, Kokar MM, Dionne R (2008) A demonstration of formal policy reasoning using an extended version of basevisor. In: IEEE workshop on policies for distributed systems and networks policy 2008. IEEE

Merriam-Webster (2011) Merriam Webster. URL http://www.merriam-webster.com/dictionary/aware

Mitola J (2000) Cognitive radio: an integrated agent architecture for software defined radio. Ph.D. thesis, KTH, Stockholm, Sweden

Mitola J (2009a) Cognitive radio policy languages. In: IEEE ICC conference. IEEE, Dresden, Germany

Mitola J (2009b) Preface. In: Fette BA (ed) Cognitive radio technology. Elsevier, Burlington

Mitola J, G Q Maguire J (1999) Cognitive radio: making software radios more personal. IEEE Pers Commun 6:13

Mizoguchi R (n.d.) On property: property vs. attribute. ISIR, Osaka University, retrieved on September 24, 2012, from http://www.ei.sanken.osaka-u.ac.jp/main/documents/OnProperty.pdf

Moskal J (2011) Interfacing a reasoner with heterogeneous self-controlling software. Ph.D. thesis, Northeastern University, Boston, USA

Moskal J, Kokar MM, Li S (2010) Interfacing a reasoner with an SDR using a thin, generic API: a GNU radio example. In: Software defined radio technical conference SDR'10. The Software Defined Radio Forum, Washington

Moy C (2010) High-level design approach for the specification of cognitive radio equipments management APIs. J Netw Syst Manag 18(1):64–96

National Communications System Technology & Standards Division (1996) Telecommunications: Glossary of telecommunication terms. Technical report, General Services Administration Information Technology Service. URL http://www.its.bldrdoc.gov/fs-1037/fs-1037c.htm, retrieved on September 24, 2012, http://www.its.bldrdoc.gov/fs-1037/fs-1037c.htm

Neel JO, Reed JH, MacKenzie AB (2009) Cognitive radio performance analysis. In: Fette BA (ed) Cognitive radio technology. Elsevier, Burlington (Chap 15)

Object Mangement Group (2007) OMG unified modeling language (OMG UML), superstructure (v2.1.2). Retrieved September 24, 2012, from http://www.omg.org/spec/UML/2.1.2/Superstructure/PDF/

Perich F, Morgan E, Ritterbush O, McHenry M, D'Itri S (2010) Efficient dynamic spectrum access implementation. In: Proceedings of the 2010 military communications conference. IEEE, pp 11427–11432

Polson J (2009) Cognitive radio: the technologies required. In: Fette BA (ed) Cognitive radio technology. Elsevier, Burlington (chap 4)

Priestnall G (2010) Introduction to variable message format (vmf). Technical report, Military Communications and Information Systems Conference MiLCIS 2010, Canberra, AU. Retrieved on Sep 24 2012, from URL http://www.milcis.com.au/milcis-final-program-2010.htm#_2.5a_ Tutorial:_Variable, retrieved on September 24, 2012, from http://www.milcis.com.au/milcis-final-program-2010.htm

Radiocommunication Study Groups (2011) Software-defined radio in the land mobile, amateur and amateur satellite services (source: Document 5a/TEMP/302(rev.1), subject: Question ITU-R 230-2/5). Technical report, International Telecommunications Union. URL http://www.itu.int/

md/dologin_md.asp?lang=en&id=R07-WP5A-C-0703!N20!MSW-E, retrieved on September 24, 2012, from http://www.itu.int/md/dologin_md.asp?lang=en&id=R07-WP5A-C-0703!N20! MSW-E

Raymer D, Strassner J, Lehtihet E, van der Meer S (2006) End-to-end model driven policy based network management. In: 7th IEEE international workshop on policies for distributed systems and networks (POLICY 2006). IEEE, pp 67–70

RDF Working Group (2004) Resource description framework (RDF). W3C, retrieved on September 24, 2012, from http://www.w3.org/RDF/

Rhyne J, Hand E, Patton S, Sperl F (2002) XML implementation of variable message format. In: IEEE software technology conference (STC). IEEE, Salt Lake City. URL http://sstc-online.org/2002/pres637.cfm

Robert M (2009) The software-defined radio as a platform for cognitive radio. In: Fette BA (ed) Cognitive radio technology. Elsevier, Burlington (Chap 3)

Rondeau TW (2007) Application of artificial intelligence to wireless communications. Ph.D. thesis, Viginia Tech

Russell SJ, Norvig P (2003) Artificial intelligence: a modern approach. Prentice Hall, Upper Saddle River

Sayin H (2003) Proposal for tactical messaging and usage of extensible markup language message text formats in the tactical command control and information systems. Master's thesis, The Middle East Technical University

Schreiber G, Dean M (2004) OWL web ontology language reference. W3C recommendation, W3C, retrieved September 24, 2012, from http://www.w3.org/TR/2004/REC-owl-ref-20040210/

Sloman A (2001) Varieties of affect and the CogAff architecture schema. In: Proceedings of the AISB'01 symposium on emotions, cognition, and affective computing. The Society for the Study of Artificial Intelligence and the Simulation of Behaviour

Smith JM (2009) Cognitive techniques: three types of network awareness. In: Fette BA (ed) Cognitive radio technology. Elsevier, Burlington (Chap 9)

Snyder J, McNair B, Edwards S, Dietrich C (2011) OSSIE: An open source software defined radio platform for education and research. In: International conference on frontiers in education: computer science and computer engineering (FECS'11). World congress in computer science, Computer engineering and applied computing. Las Vegas, NV

Stewart D (2009) Military and public safety dsa policy use case. Technical report, IEEE SCC 41 Working Group, IEEE

Studer R, Benjamins VR, Fensel D (1998) Knowledge engineering: principles and methods. Data Knowl Eng 25(1–2):161–197

Stuntebeck E, O'Shea T, Hecker J, Clancy T (2006) Architecture for an open-source cognitive radio. In: Proceedings of the SDR forum technical conference 2006, SDRF

Tanenbaum AS (2002) Computer networks, 4th edn. Prentice Hall PTR, Upper Saddle River

Taylor JH, Sayda AF (2005) An intelligent architecture for integrated control and asset management for industrial processes. In: Proceedings of the 2005 IEEE international symposium on intelligent control. IEEE, pp 1397–1404

Thompson HS, Maloney M, Beech D, Mendelsohn N (2004) XML schema part 1: structures, 2nd edn. W3C recommendation, W3C, retrieved September 24, 2012, from http://www.w3.org/TR/2004/REC-xmlschema-1-20041028/

Transceiver Interface Task Working Group (2009) Transceiver facility specification (SDRF-08-s-0008-v1.0.0). Technical report, The Software Defined Radio Forum, retrieved on September 24, 2012, from http://www.sdrforum.org/pages/documentLibrary/documents/SDRF-08-S-0008-V1_0_0_Transceiver_Facility_Specification.pdf

Tsui E (2004) What are adaptive, cognitive radios? Technical report, Berkeley Wireless Research Center, retrieved on September 24, 2012, from http://bwrc.eecs.berkeley.edu/php/pubs/pubs.php/649/ACR%20E%20Tsui.ppt

US Joint Program Executive Office (2011) Software communications architecture (SCA). Retrieved September 24, 2012, from http://www.public.navy.mil/jpeojtrs/sca/Pages/default.aspx

Vasudevan V (1998) Comparing agent communication languages. Technical report, OBJS Technical Note, URL http://act-r.psy.cmu.edu/people/douglass/Douglass/Douglass/Agents/ TopicPapers/KQML/9807-comparing-ACLs.pdf, retrieved on September 24,2012, http:// act-r.psy.cmu.edu/people/douglass/Douglass/Douglass/Agents/TopicPapers/KQML/9807-comparing-ACLs.pdf

W3C (2006) Sparql query language for rdf. Retrieved September 24, 2012, from http://www.w3. org/TR/2006/CR-rdf-sparql-query-20060406/

W3C OWL Working Group (2009) OWL 2 web ontology language document overview. Technical report, W3C, retrieved on September 24, 2012, from http://www.w3.org/TR/2009/REC-owl2-overview-20091027/

Walmsley P, Fallside DC (2004) XML schema part 0: Primer second edition. W3C recommendation, W3C, retrieved September 24, 2012, from http://www.w3.org/TR/2004/REC-xmlschema-0-20041028/

Wang J, Brady D, Kokar M, Lechowicz L (2003) The use of ontologies for the self-awareness of the communication nodes. In: Software defined radio technical conference SDR'03

Wang J, Kokar M, Baclawski K, Brady D (2004) Achieving self-awareness of sdr nodes through ontology-based reasoning and reflection. In: Software defined radio technical conference SDR'04. The Software Defined Radio Forum, Phoenix

Wilkins D, Denker G, Stehr MO, Elenius D, Senanayake R, Talcott C (2007) Policy-based cognitive radios. IEEE Wireless Commun 14(4):41–46

Willink ED (2000) Fm3tr decompostion (p6957-11-005). Technical report, DERA

Willmott S, Faltings B, andS Macho-Gonzalez MC, Belakhdar O, Torrens M (1999) Constraint choice language (CCL) language specification v2.01. Technical report, Universitat Politecnica de Catalunya

Xuan P, Lesser V, Zilberstein S (2001) Communication decisions in multi-agent cooperation: model and experiments. In: The fifth international conference on autonomous agents. Association for Computing Machinery, Montreal

Index

S. Li and M.M. Kokar, *Flexible Adaptation in Cognitive Radios*,
Analog Circuits and Signal Processing, DOI 10.1007/978-1-4614-0968-7,
© Springer Science+Business Media, LLC 2013